STP 988

Functional Testing of Aquatic Biota for Estimating Hazards of Chemicals

John Cairns, Jr., and James R. Pratt, editors

ASTM
1916 Race Street
Philadelphia, PA 19103

Library of Congress Cataloging-in-Publication Data

Functional testing of aquatic biota for estimating hazards of
  chemicals/John Cairns, Jr., and J. R. Pratt, editors.
    (STP; 988)
    "Papers presented at the Symposium on Functional Testing for
  Hazard Evaluation, which was held in Bal Harbour, Florida, on 6–7
  Nov. 1986. This symposium was sponsored by ASTM Committee D-19 on
  Water"—Foreword.
    Includes bibliographies and indexes.
    "ASTM publication code number (PCN) 04-988000-16."
    ISBN 0-8031-1165-7
    1. Water quality bioassay—Congresses. 2. Water—Pollution—
  Toxicology—Congresses. 3. Water—Pollution—Environmental
  aspects—Congresses. I. Cairns, John, Jr., 1923-   . II. Pratt, J. R.
  (James Richard), 1950-   . III. Symposium on Functional Testing for
  Hazard Evaluation (1986: Bal Harbour, Fla.). IV. American Society
  for Testing and Materials. Committee D-19 on Water. V. Series:
  ASTM special technical publication; 988.
  QH96.8.B5F86   1988
  628.1′61—dc19   88-31528
                                                              CIP

Copyright © by AMERICAN SOCIETY FOR TESTING AND MATERIALS 1988

Peer Review Policy

Each paper published in this volume was evaluated by three peer reviewers. The authors addressed all of the reviewers' comments to the satisfaction of both the technical editor(s) and the ASTM Committee on Publications.

The quality of the papers in this publication reflects not only the obvious efforts of the authors and the technical editor(s), but also the work of these peer reviewers. The ASTM Committee on Publications acknowledges with appreciation their dedication and contribution of time and effort on behalf of ASTM.

Printed in Ann Arbor, MI
January 1989

Foreword

This publication, *Functional Testing of Aquatic Biota for Estimating Hazards of Chemicals*, contains papers presented at the Symposium on Functional Testing for Hazard Evaluation, which was held in Bal Harbour, Florida, on 6–7 Nov. 1986. The symposium was sponsored by ASTM Committee D-19 on Water. John Cairns, Jr., Virginia Polytechnic Institute and State University, presided as symposium chairman. Dr. Cairns and James R. Pratt, Pennsylvania State University, were the editors of this publication.

Contents

Introduction

When a physician gives an annual physical, he or she generally makes the major judgment of the patient's health on the basis of functional attributes: blood pressure, heart rate, and the performance of the kidneys as demonstrated by the quality of the urine. Structural attributes, such as skin and eye condition, structural features demonstrated by X-rays, and other charcteristics are also given attention. When environmental health is being determined, structural attributes based on "critter counting" are generally the most important determinants. Functional attributes of natural systems, such as energy transfer, nutrient spiraling, or rate of carbon fixation, are generally not nearly as prominent in regulatory measures to protect indigenous biota. This indicates that, where ecosystem health is concerned, we are more interested in the condition than in performance. We want to be sure that threatened and endangered species remain. The Europeans, with their saprobian system, extrapolate from the presence of certain organisms to the overall ecosystem condition. However, a species may be present, but only in marginal condition or functioning poorly. If enough species are functioning poorly, the ecosystem function will deteriorate as well. Moreover, performance is an integrating function, which is to say that, if any one of a number of components is malfunctioning, the entire system may perform poorly, whereas the presence or absence of species is a less integrated assessment. Of course, if the system is malfunctioning, one must make a detailed examination to determine why, which is precisely what a physician does if some of the characteristics assessed during the annual physical are outside normal boundaries. This means that, from a cost-effectiveness standpoint, if everything is going well and one measures a single integrated function that reflects the well-being of a variety of components, no further testing is necessary. If it is necessary, naturally the cost has increased.

Although this is written as if structure and function were dichotomous choices, it was done for illustration rather than as a recommendation that one approach be substituted for the other. My own feeling is that multiple lines of evidence are needed in a desirable management and regulatory approach when assessing the well-being of a complex system. The point is that, while we have structural and functional components in good balance for annual physicals of human beings, we have not yet achieved what appears to be a desirable balance for assessment of environmental or ecological health. Part of the reason is that ecologists have focused on species presence or absence rather than on ecosystem function, although this has changed in recent years. Initially, the methodology for structural assessment was more readily available than that for functional assessment of ecosystems.

The purpose of this book is to provide some illustrations of various types of functional measurements and how they might be used. This is by no means an exhaustive compilation—which would require a volume an order of magnitude larger—but rather an illustrative selection so that readers of ASTM publications can see what is available and how it might be used. It is quite likely that, with the advent of mesocosm testing for predicting the effects of pesticides on ecosystems in the United States, functional attributes will receive much more attention than they have in the past. These tests are strikingly more expensive than single-species toxicity tests involving lethality, reproductive success, and the like, and, therefore, any integrating function that can increase the information value while simultaneously keeping the cost constant or reducing the cost is well worth considering. This is a newly developing field and is still in the exploratory stage. There is no large body of standard methods presently available. However, some of the methods in this publication may, with modifications, be quite suitable for utilization as standard methods after a sufficient number of professionals have used them and their weaknesses and strengths have become better known.

John Cairns, Jr.

University Center for Environmental and Hazardous Materials Studies, Virginia Polytechnic Institute and State University, Blacksburg, VA 24061; symposium chairman and editor.

James R. Pratt

School of Forest Resources, Pennsylvania State University, University Park, PA 16802; editor.

John F. McCarthy[1] and Steven M. Bartell[1]

How the Trophic Status of a Community Can Alter the Bioavailability and Toxic Effects of Contaminants

REFERENCE: McCarthy, J. F. and Bartell, S. M., **"How the Trophic Status of a Community Can Alter the Bioavailability and Toxic Effects of Contaminants,"** *Functional Testing of Aquatic Biota for Estimating Hazards of Chemicals, ASTM STP 988,* J. Cairns, Jr., and J. R. Pratt, Eds., American Society for Testing and Materials, Philadelphia, 1988, pp. 3-16.

ABSTRACT: Binding of hydrophobic organic and metal contaminants to particulate organic matter (POM) and dissolved organic matter (DOM) in aquatic systems affects the availability of contaminants and their subsequent dose to biota. Eutrophic systems that contain high levels of organic sorbents will have lower concentrations of freely dissolved, readily available toxicant, thus reducing the exposure of the biota. In more oligotrophic systems with lower levels of sorbents, toxic exposure may be greater. These interactions suggest a relationship between the trophic state of an ecosystem and its susceptibility to toxic effects. Studies on the binding of organic contaminants to POM and DOM and its effects on toxicant accumulation are reviewed. The role of system productivity in producing adverse effects is explored by computer simulations, using a combined fate and effects model to evaluate the impacts of naphthalene on the production dynamics and contaminant body burden in different model populations when the concentration of POM varied from 0 to 10 mg of carbon per litre. Higher levels of POM decreased body burdens and moderated the reduction in productivity resulting from the exposure to naphthalene. Any interpretation of functional tests used to evaluate hazardous substances should consider the interaction between the trophic state of the system and the potential dose of toxicant available to biota.

KEY WORDS: hazard evaluation, trophic status, bioaccumulation, sorption, contaminants, dissolved organic matter, humic acid, fate and effect, ecosystem models

Functional tests for assessing impacts due to the release of effluents or chemicals into an aquatic system have obvious advantages for providing ecologically relevant information on toxic effects. However, the whole-system complexity that makes these approaches attractive can also complicate interpretation of the results. This paper will address the role of the trophic state of an ecosystem in the bioavailability of contaminants. An argument will be made that systems with different levels of productivity will respond very differently to the same loading of contaminant. Biota in a eutrophic system will not receive the same dose of contaminant as biota in an oligotrophic system because more of the contaminant will bind to the higher levels of particulate and dissolved sorbents in the eutrophic system, thereby becoming less available to biota and exerting less toxic effect.

An illustrative example is a case in which the same effluent is delivered at the same loading rate to two systems, one eutrophic and one oligotrophic (Fig. 1). A set of functional responses (such as a change in population growth or carbon fixation) is measured in both systems with the intention of determining the hazards posed by the release of the effluent to aquatic systems. If the tests indicate that productivity or nutrient uptake is more affected in the oligotrophic system

---

[1]Research staff members, Environmental Sciences Division, Oak Ridge National Laboratory, Oak Ridge, TN 37831.

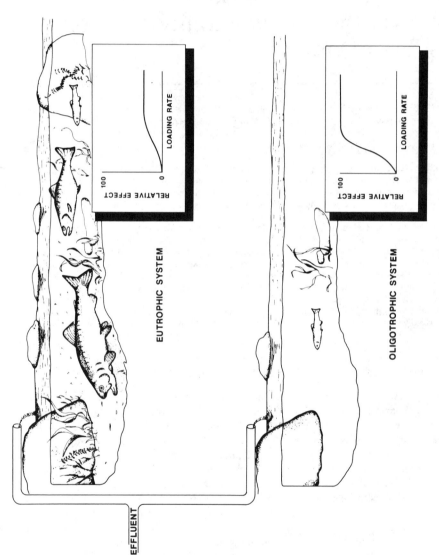

FIG. 1—*An illustrative example of the impact of loading the same toxicant in two ecosystems with different levels of productivity. The eutrophic system shows relatively less effect than the oligotrophic system in response to identical loading rates of the same effluent.*

than in the eutrophic system, this result may be interpreted as indicating that the effluent entering the oligotrophic system is more toxic or hazardous than the effluent entering the eutrophic system. In fact, it is the same effluent, but the trophic state of the ecosystem altered the amount of toxicant available to exert a deleterious effect.

In this paper, the authors first discuss the role of physicochemical partitioning of contaminants on the bioavailability of the chemical. We then present the results of a computer simulation using a model of contaminant fate and effects to illustrate how a functional end point, the productivity of different model populations, can vary as a result of differences in the concentration of sorbent in the system. Throughout, we use organic compounds as model contaminants, but metals also bind to particulate and dissolved sorbents, and their toxicity might vary in a manner analogous to that for the organics.

Physicochemical Partitioning in the Aquatic Environment

The affinity of hydrophobic organic contaminants for binding to sediment and suspended particles in aquatic systems is directly related to the hydrophobicity of the contaminant and the organic carbon content of the particles [1-4]. The hydrophobicity of a contaminant can be quantified as its octanol-water partition coefficient, K_{ow}. The affinity of a chemical for binding to a sorbent can be quantified as an association coefficient, K_p, defined here as the ratio of the amount of compound bound per gram of particles and the amount dissolved per millilitre of water when the system is at equilibrium. [This is demonstrated in Eq 1, where C_p is the amount of contaminant (in picomoles per millilitre) bound to particles; C_d is the amount of dissolved contaminant (in picomoles per millilitre); and P is the mass of particles (in grams per millilitre).] Different sediments or particles exhibit different affinities for binding the same contaminant, although the differences can be normalized by correcting for the organic carbon content of the sorbing particle, f_{oc} (Eq 2). This carbon-referenced association coefficient, K_{oc}, has been related to the K_{ow} of a contaminant through several regression equations [3-5], such as those indicated here (Eqs 3a and 3b)

$$K_p = \frac{C_p}{C_d \times P} \tag{1}$$

$$K_{oc} = \frac{K_p}{f_{oc}} \tag{2}$$

$$\log K_{oc} = \log K_{ow} - 0.317 \tag{3a}$$
(5 polycyclic aromatic hydrocarbons; Ref 3)

$$\log K_{oc} = 0.72 \log K_{ow} + 0.49 \tag{3b}$$
(methylated and halogenated benzenes; Ref 5)

Other important, but less obvious, sorbents for hydrophobic organic contaminants are naturally occurring dissolved organic macromolecules and colloids, collectively referred to here as dissolved organic matter (DOM). As with particulate sorbents, there is a log linear relationship between the K_{ow} of hydrophobic contaminants and their affinity for binding to DOM [6-9]. Because of the difficulties of quantifying the amount of contaminant bound to DOM, its role in altering the fate and bioavailability of contaminants has often been ignored. Contaminant fate is usually analyzed in terms of partitioning between two phases: a particulate phase, C_p, and a nonparticulate phase, C_{np}. In fact, the presence of DOM forms a third phase; the nonparticulate phase actually consists of contaminant which is truly dissolved, C_d, and contaminant bound to DOM. [This is demonstrated in Eqs 5 and 6, where C_{DOM} is the amount of contaminant bound to DOM (in picomoles per millilitre), and DOM is the concentration of DOM (in grams of carbon per millilitre).] To account for the role of DOM, partitioning requires analysis in terms of two interactions: (1) binding of C_d to particles and (2) binding of C_d to DOM. The apparent association coefficient, K_{app}, describing the partitioning between the particulate and nonparticulate phases, can be described in terms of these independent interactions [10]

$$K_1 = K_p \text{ for particles} = \frac{C_p}{C_d \times P} \tag{4}$$

$$K_2 = K_p \text{ for DOM} = \frac{C_{DOM}}{C_d \times DOM} \tag{5}$$

$$K_{app} = \frac{C_p}{C_{np} \times P} = \frac{C_p}{(C_d + C_{DOM}) \times P} \tag{6a}$$

$$K_{app} = \frac{K_1}{1 + K_2 \times DOM} \tag{6b}$$

The amount of contaminant associated with particles will decrease as the concentration of DOM increases (Eq 6b). This partitioning can be conceptualized as the distribution of the hydrophobic contaminant between a polar aqueous environment and two nonpolar phases: (1) the particulate organic matter (POM), which is $P \times f_{oc}$ (in grams of carbon per millilitre), and (2) the DOM. As with sediments, the affinity of a chemical for binding to each can be approximated as the K_{oc} of the contaminant [10].

Bioavailability of Contaminants

The physicochemical partitioning of contaminants must be understood to evaluate the potential for accumulation of the chemical by biota. Association of a contaminant with DOM or POM significantly changes the bioavailability of the toxicant and, thus, the potential dose experienced by the organisms. Alteration of the dose would be expected to affect the magnitude of any toxic effect on individuals, populations, or communities.

For example, it has been experimentally demonstrated that the presence of yeast cells significantly decreases the accumulation of polycyclic aromatic hydrocarbons (PAHs), anthracene, and benzo(a)pyrene by Daphnia magna (Fig. 2) [11]. The suspended yeast particles bound a fraction of the total polycyclic aromatic hydrocarbon (PAH) in the experimental system, thus decreasing the amount of contaminant remaining in the dissolved phase. Analysis of the uptake kinetics indicated that the PAH dissolved in the water was much more rapidly taken up by Daphnia magna than the PAH bound to the suspended particles. Other researchers have also shown that sediment- or particle-bound contaminants are much less bioavailable than contaminants dissolved in the water [12-14].

Binding to DOM can have an even greater effect on the accumulation of contaminants since a chemical bound to DOM appears to be unavailable for uptake by aquatic organisms. Figure 3 illustrates the effect of DOM on the accumulation of a series of PAHs by D. magna. Analysis of these data using either kinetic or steady-state approaches demonstrated that PAH bound to DOM is not available to D. magna [15]. Similar results were reported for other aquatic invertebrates and for fish [16-18]. The bioavailability of PAH bound to DOM was examined in more detail using a fish metabolic chamber designed to measure the efficiency with which contaminants are extracted from water by the gills of rainbow trout [19]. The presence of DOM reduced the uptake of PAH; the extent of the reduction in uptake efficiency was equal to the fraction of PAH that was removed from aqueous solution by its association with DOM. While uptake of the

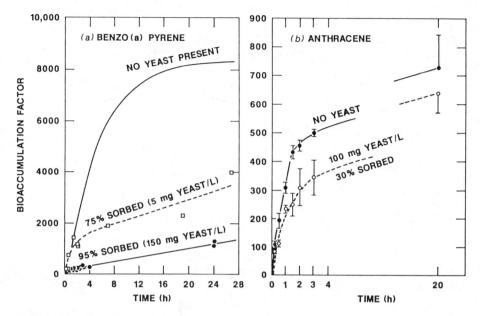

FIG. 2—*The effect of suspended yeast particles on the accumulation of* (a) *benzo(a)pyrene and* (b) *anthracene by* Daphnia magna. *Groups of five animals were maintained in 100-mL jars containing carbon-14 labeled* (^{14}C) *PAH, with or without yeast. The jars were slowly rotated to prevent settling of the particles. At various times, replicate jars were sampled and the body burden of PAH in the* D. magna *was determined by liquid scintillation counting* [11]. *The figure is reproduced by permission of Springer-Verlag, New York.*

total amount of PAH decreased, the uptake efficiency of the fraction of total PAH remaining truly dissolved in the water, measured using equilibrium dialysis [6], remained constant, regardless of the concentration of DOM in the water. These results demonstrate that only dissolved PAH was translocated across the fish gills [20].

The environmental significance of this interaction is illustrated in Fig. 4. The reduction in bioconcentration is greatest for the more hydrophobic compounds, which have the greatest affinity for binding to DOM. These compounds would include the more environmentally persistent chemicals—for example, PAHs, polychlorinated biphenyls, dioxin, Kepone, mirex and others—which pose the greatest health risks as well as provide the greatest potential for accumulation in the food chain.

Ecosystem Productivity as a Determinant of Partitioning and Bioavailability

Ecosystems can differ widely in their productivity and thus their concentrations of suspended POM and DOM. Oligotrophic systems will have low levels of organic matter in either a particulate or dissolved form. Inputs of nutrients or allochthonous organic matter stimulate productivity and provide, either directly or indirectly, a rich source of particulate and dissolved sorbents in eutrophic systems. The dosage of contaminant to which biota are exposed will be significantly reduced in eutrophic systems because of the interaction of the chemicals with these sorbents. The presence of organic sorbents will reduce the fraction of the total contaminant in the freely dissolved, most readily bioavailable form. The presence of DOM will also decrease the amount of contaminant that will sorb to particles and potentially enter the food chain through ingestion by zooplankton, detrivores, or benthic organisms. The fraction of the total amount of contami-

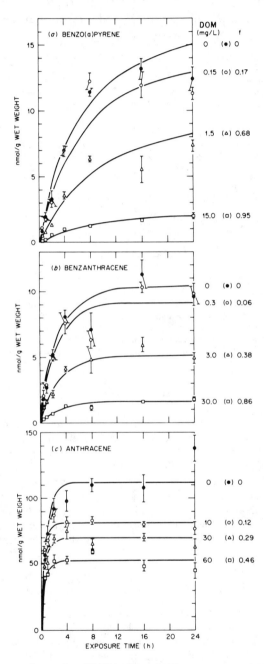

FIG. 3—*The time course for uptake of PAH is shown for* D. magna *exposed to* (a) *benzo(a)pyrene,* (b) *benzanthracene, and* (c) *anthracene in the presence of different concentrations of DOM. The presence of 15 to 60 mg carbon/L as DOM did not significantly decrease the accumulation of naphthalene (data not shown). The mean concentrations in the animals (plus the standard errors of four replicates) are plotted. The concentration of DOM (in milligrams of carbon per litre) and the fraction of PAH bound to the DOM,* f, *are indicated* [14]. *The figure is reproduced by permission of Elsevier Publishing Co., New York.*

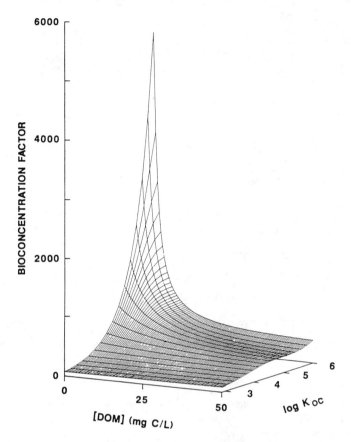

FIG. 4—*Structure-activity relationship between the* K_{oc}, *DOM concentration, and bioconcentration factor (BCF) of PAHs. The left edge indicates the observed relationship between* K_{oc} *of a series of PAHs and their bioaccumulation by* D. magna *when no DOM is present. For a less hydrophobic PAH, such as naphthalene (front edge of the figure, log* $K_{oc} = 3$*), the BCF is low in the absence of DOM, and the presence of DOM has relatively little effect. However, for a very hydrophobic PAH like benzo(a)pyrene (back edge of the figure, log* $K_{oc} = 6$*), even small amounts of DOM bind a significant fraction of the PAH and decrease bioaccumulation [11]. The figure is reproduced by permission of Elsevier Publishing Co.*

nant entering a system that will be either freely dissolved, f_d, or bound to POM f_p, can be approximated by [*10*]

$$f_d = \frac{1}{1 + K_{oc} \times \text{POM} + K_{oc} \times \text{DOM}} \tag{7}$$

$$f_p = \frac{K_{oc} \times \text{POM}}{1 + K_{oc} \times \text{POM} + K_{oc} \times \text{DOM}} \tag{8}$$

As the total amount of organic sorbent (POM + DOM) increases, the potential dose to biota from the bioavailable aqueous phase of the contaminant decreases. The reduction in potential dose from this route of exposure can be calculated as $(1 - f_d)$ and is plotted in Fig. 5 as a function of the concentration of total organic sorbent and of the K_{oc} of contaminants. While the

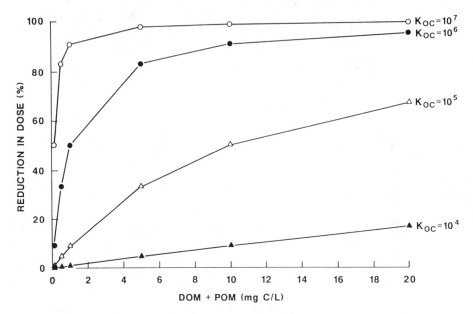

FIG. 5—*The reduction in the dose of contaminant taken up through fish gills is illustrated. The relationship is based on the observation that only freely dissolved contaminant crosses the gill membrane. The reduction in uptake is greater for more hydrophobic compounds, which have a higher affinity for binding to DOM.*

reduction in dose is greatest for the most hydrophobic compounds, even relatively small reductions in dose could substantially decrease an organism's exposure to levels below critical thresholds for acute or chronic effects.

The effect of DOM on the binding of contaminants to POM varies with the relative concentration of the two sorbents and the hydrophobicity of the contaminant. Figure 6 illustrates the change in the fraction of the total contaminant in the system which will be bound to a low POM loading (simulating an oligotrophic ecosystem, Fig. 6*a*) or a high level of POM (a eutrophic system, Fig. 6*b*). The extent of binding to suspended particles increases dramatically in the more eutrophic case and DOM exerts quite different effects in the two systems. In the absence of DOM, binding increases as the K_{oc} of the contaminant increases; more contaminant is bound at higher loadings of POM. Even low levels of DOM can reduce or eliminate binding to particles when POM concentrations are low (Fig. 6). DOM has less effect when POM levels are high (Fig. 6).

These examples illustrate the potential nature and magnitude of the effects of physicochemical interactions on the potential dose of toxicant incorporated by biota. Far more subtle interactions are possible when the complexities of trophic interactions and toxicological effects are considered.

Combined Fate and Effects Model Simulations

The relationships, data, and examples presented thus far were derived under laboratory conditions of constant temperature and light for single populations of *Daphnia* or fish. Bartell et al. [21] describe a model that simulates the fate and effects of PAHs under seasonal light and temperature regimes involving nine interacting populations of aquatic plants and animals. This model was used to test the null hypothesis that the accumulation and toxic effects of naphthalene by populations in aquatic systems are independent of the concentrations of POM.

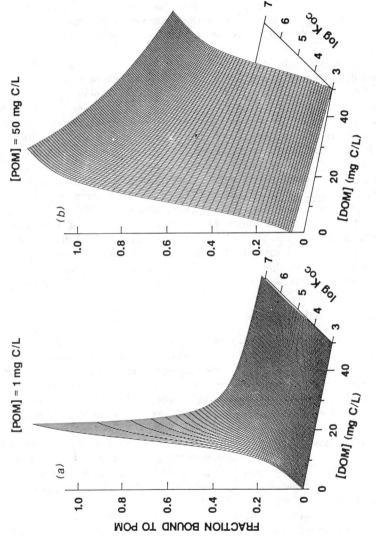

FIG. 6—*The effect of the presence of different concentrations of DOM on the fraction of contaminant bound to two concentrations of suspended POM: (a) 1 mg carbon/L as POM, and (b) 50 mg carbon/L as POM.*

Model Description

The combined fates and effects model (CFEM) consists of coupled differential equations that describe temporal changes in the concentration of dissolved PAH and the biomass and body burden of PAH in three populations of primary producers and six populations of consumers. The model processes include volatilization [22], photolysis [23,24], and sorption [3,4] of dissolved PAH, as well as direct uptake and depuration of PAH by the model biota [25]. Bartell et al. [21] detail the translation of accumulated PAH to decreases in growth rates of the model populations. The current model integrates these dynamics under a single square metre of a completely mixed 10-m-deep water column. Values for kinetics of naphthalene uptake and depuration, as well as toxicity data used to derive population-specific dose-response functions, are listed in Bartell et al. [21].

The CFEM can also address the relative accumulation of dissolved PAH and PAH contributed by contaminated prey populations. Given this study's emphasis on organic matter as a substrate competing with biota for sorption of dissolved compound, we omitted food-web accumulation by defining the assimilation of ingested naphthalene as equal to zero. The lack of data quantifying such assimilation for organisms representative of the model populations further justified this omission. It should also be noted that naphthalene has a relatively low affinity, in comparison with other PAHs, for binding to organic sorbents or for bioaccumulation. Effects due to sorption of contaminants would be expected to be greater for more hydrophobic PAHs.

To evaluate the null hypothesis concerning the effects of suspended POM on accumulation of napthalene, simulations were performed with the CFEM in which concentrations of POM were defined as 0, 0.1, 1.0, and 10.0 mg of carbon per litre. A constant daily input of dissolved naphthalene of 0.32 mg/L (1.0% saturation solution) was used in all simulations. The daily production dynamics and body burden of naphthalene were recorded in relation to the POM concentration for model days 170 through 240. This period constitutes a major portion of the summer period of biomass production by the model populations. Simulated concentrations of dissolved naphthalene were also recorded.

Simulation Results

The percentage change in biomass production varied in relation to the modeled concentration of POM under conditions of constant inputs of naphthalene (Table 1). The model results emphasized decreased toxic effects, measured as percentage changes in the total biomass production during summer, with increased concentrations of POM. The simulation results also reflect the population-specific growth rates and kinetics for uptake and depuration of naphthalene, in addition to the differing sensitivity of model populations to naphthalene [21]. For periphyton, macrophytes, and detrivorous fish, the decrease in toxic effects in relation to increased POM concentrations was relatively minor, on the order of 10% or less. Phytoplankton, zooplankton, benthic invertebrates (for example, clams and crayfish), and omnivorous fish, alternatively, showed 15 to 30% reductions in toxic effects with increased POM. The model results for zooplankton and bacteria were quantitatively different from those for other populations: at POM concentrations of 0 and 0.1 mg carbon/L, toxic effects were calculated, while at POM concentrations of 1.0 and 10.0 mg carbon/L, the utilization of POM as a food source with subsequent stimulation of growth exceeded the toxic effects of naphthalene, and productivity of these populations actually increased. For benthic insects, the population least sensitive to naphthalene [21], productivity was enhanced even in the absence of added POM. Regardless of the POM concentration, mortality losses from other model populations entered a settled detritus pool that was fed upon by the modeled insects.

The results summarized in Table 1 do not support the null hypothesis. If it is assumed that concentrations of POM are positively correlated with the trophic status of the system, these results imply that eutrophic systems will exhibit less response than oligotrophic systems to low-level chronic additions of naphthalene and perhaps other hydrophobic toxic chemicals. The

TABLE 1—*Percentage decrease in net production of biomass as a function of the concentration of suspended particulate organic matter (POM) and a constant input of naphthalene (0.32 mg/L/day).*[a,b]

Model Population	Suspended Particulate Organic Matter, mg C/L			
	0	0.1	1.0	10
Phytoplankton	20.9	20.1	14.7	6.9
Periphyton	36.9	36.8	34.6	30.8
Macrophytes	53.8	53.8	52.5	50.6
Zooplankton	2.5	0.3	+8.1[b]	+23.9
Benthic insects	+23.9	+25.5	+67.2	+660.1
Benthic invertebrates	41.8	41.5	34.7	11.8
Bacteria	59.9	27.3	+121.5	+963.5
Omnivorous fish	42.8	42.5	37.4	16.1
Detrivorous fish	67.2	67.1	64.9	61.5

[a]The reference simulation had no naphthalene and no added POM.
[b]Note that + denotes an increase in production.

differing uptake, depuration, and sensitivity of populations to the toxicant may release populations from competitive or predator/prey constraints on growth [26,27]. Such a release might offset the direct effects of the toxicant on population growth rate, producing increased biomass, in comparison with growth in the absence of the toxicant.

The relationship between increased POM and decreased toxicity was posited to result from the effective removal of naphthalene from solution through sorption. Increased POM would simply reduce the potential for accumulation of the compound, reduce the dosage, and subsequently decrease the toxic effects. The results of the simulations were consistent with this explanation. Because *Daphnia* have been the focus of experimental work leading to the supposition just described, model results for the zooplankton population are presented. Similar results were observed for other model populations. Concentrations of dissolved naphthalene decreased in relation to added POM (Fig. 7). The POM addition of 0.1 mg carbon/L/day decreased dissolved naphthalene by as much as 23% relative to the no-POM simulation. The 1.0 and 10.0-mg carbon/L/day POM simulations decreased naphthalene in solution by as much as 53%. The reduced concentrations of dissolved naphthalene were associated with comparative decreases in the modeled naphthalene concentration in the zooplankton (Fig. 8). The 0.1 mg carbon/L POM simulation, representative of relatively unproductive aquatic systems, produced minimal changes in body burden relative to the zero POM simulation; however, the higher POM additions, indicative of more productive systems, reduced body burden by as much as 14 to 45% of that expected in the absence of POM.

The implications of our incomplete and imperfect understanding of the environmental behavior of naphthalene were summarized as an operational hypothesis, the model. Despite the necessary simplifications and assumptions given our current quantitative understanding, the model results are consistent with logical predictions based on the results of experiments performed under more constrained laboratory conditions. Forecasts of the effects of hydrophobic toxic chemicals added to natural systems should include consideration of their trophic status.

Summary and Conclusions

Some of the physicochemical interactions of contaminants with organic sorbents in ecosystems have been reviewed to emphasize that these natural components can alter the bioavailability of contaminants and can modulate the dose of toxicant incorporated by biota. From the

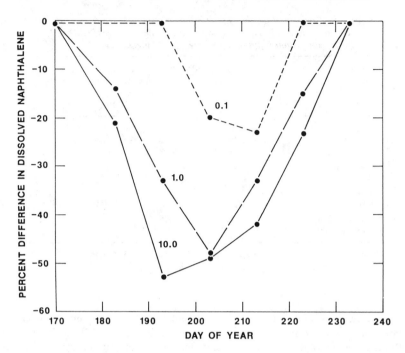

FIG. 7—*The results of CFEM simulations indicating the percentage change in the concentration of dissolved naphthalene in the presence of POM (0.1, 1.0, and 10 mg carbon/L), relative to the dissolved concentration predicted when no POM is present.*

fundamental dose-response principle in toxicology, factors modifying the uptake of a contaminant would be expected to alter the nature and extent of adverse effects at the individual, population, or community level. The computer simulations provided an example of how simply changing the amount of POM in a model ecosystem may significantly alter the effects of identical loading of the same toxicant in terms of both the accumulation of toxicant by organisms and the productivity of the ecosystem.

This exercise is cautionary. Care must be taken in using functional tests of ecosystem performance to evaluate the potential toxicity and hazards posed by the release of chemical effluents. Without considering the complex toxicological and ecological interactions underlying such testing, even seemingly straightforward results can lead to quite erroneous conclusions. The potential hazard of any effluent must also be considered in terms of the ecosystem into which it will be discharged. An effluent entering a productive stream or lake may pose little hazard, but the same material may have a very deleterious effect on a more oligotrophic system. The answer to the questions of "How safe is safe?" and "How clean is clean?" may be "It depends." The answer will be determined, at least in part, by the trophic status of the receiving ecosystem.

Acknowledgments

The authors are grateful to G. W. Suter for valuable discussions and for suggesting the theme for this manuscript. We thank him, G. R. Southworth, and M. C. Black for reviewing this manuscript. This work was sponsored jointly by the Oak Ridge Y-12 Plant, Department of Environmental Management, Health, Safety, Environment and Accountability Division, and by the U.S. Environmental Protection Agency under Interagency Agreement DW89930690-01-

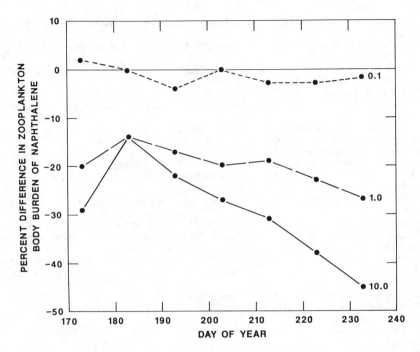

FIG. 8—*The results of CFEM simulations illustrating the difference in accumulated body burden in zooplankton when different levels of POM are present, relative to body burdens predicted in the absence of POM.*

0 with the U.S. Department of Energy. The Oak Ridge Y-12 Plant and the Oak Ridge National Laboratory are operated by Marietta Energy Systems, Inc., under Contract No. DE-AC05-840R21400 with the U.S. Department of Energy. This is publication No. 3001 of the Environmental Sciences Division, Oak Ridge National Laboratory.

References

[1] Karickhoff, S. W., Brown, D. S., and Scott, T. A., *Water Research,* Vol. 13, 1979, pp. 241–248.
[2] Means, J. C., Wood, S. G., Hassett, J. J., and Branwart, W. L., *Environmental Science and Technology,* Vol. 14, 1980, pp. 1524–1528.
[3] Karickhoff, S. W., *Chemosphere,* Vol. 10, 1981, pp. 833–846.
[4] Karickhoff, S. W., *Journal of Hydraulic Engineering,* Vol. 110, 1984, pp. 707–735.
[5] Schwartzenbach, R. P. and Westall, J., *Environmental Science and Technology,* Vol. 15, 1981, pp. 1360–1367.
[6] McCarthy, J. F. and Jimenez, B. D., *Environmental Science and Technology,* Vol. 19, 1985, pp. 1072–1076.
[7] Carter, C. W. and Suffett, I. H., *Environmental Science and Technology,* Vol. 16, 1982, pp. 735–740.
[8] Landrum, P. F., Nihart, S. R., Eadie, B. J., and Gardner, W. S., *Environmental Science and Technology,* Vol. 18, 1984, pp. 187–192.
[9] Wijayaratne, R. D. and Means, J. C., *Environmental Science and Technology,* Vol. 18, 1984, pp. 121–123.
[10] McCarthy, J. F. and Black, M. C., "Partitioning Between Dissolved Organic Macromolecules and Suspended Particles: Effects on Bioavailability and Transport of Hydrophobic Organic Chemicals in Aquatic Systems," *Aquatic Toxicology and Hazard Assessment: Tenth Volume, ASTM STP 971,* American Society for Testing and Materials, Philadelphia, 1988, pp. 233–246.
[11] McCarthy, J. F., *Archives of Environmental Contamination and Toxicology,* Vol. 12, 1983, pp. 559–568.

[12] Landrum, P. F. and Scavia, D., *Canadian Journal of Fisheries and Aquatic Sciences*, Vol. 40, 1983, pp. 298-305.
[13] Muir, D. C. G., Townsend, B. E., and Lockhart, W. L., *Environmental Science and Technology*, Vol. 17, 1983, pp. 269-281.
[14] Hall, W. S., Dickson, K. L., Saleh, F. Y., and Rodgers, J. H., *Archives of Environmental Contamination and Toxicology*, Vol. 15, 1986, pp. 529-534.
[15] McCarthy, J. F., Jimenez, B. D., and Barbee, T., *Aquatic Toxicology*, Vol. 7, 1985, pp. 15-24.
[16] McCarthy, J. F., and Jimenez, B. D., *Environmental Toxicology and Chemistry*, Vol. 4, 1985, pp. 511-521.
[17] Leversee, G. J., Landrum, P. F., Giesy, J. P., and Fannin, T., *Canadian Journal of Fisheries and Aquatic Sciences*, Vol. 40, Supplement 2, 1983, pp. 63-69.
[18] Carlberg, G. E., Martinsen, K., Kringstad, A., Gjessing, E., Grande, M., Kallquist, T., and Skare, J. U., *Archives of Environmental Contamination and Toxicology*, Vol. 15, 1986, pp. 811-816.
[19] McKim, J. M. and Goeden, H. M., *Comparative Biochemistry and Physiology*, Vol. 72C, 1982, pp. 65-74.
[20] Black, M. C. and McCarthy, J. F., *Environmental Toxicology and Chemistry*, Vol. 7, 1988, pp. 593-600.
[21] Bartell, S. M., Gardner, R. H., and O'Neill, R. V., "An Integrated Fates and Effects Model for Estimation of Risk in Aquatic Systems," *Aquatic Toxicology and Hazard Assessment: Tenth Volume, ASTM STP 971*, American Society for Testing and Materials, Philadelphia, 1988, pp. 261-274.
[22] Southworth, G. R., "The Role of Volatilization in Removing Polycyclic Aromatic Hydrocarbons from Aquatic Environments," *Bulletin of Environmental Contamination and Toxicology*, Vol. 21, 1986, pp. 507-514.
[23] Zepp, R. G. and Cline, D. M., "Rates of Direct Photolysis in Aquatic Environments," *Environmental Science and Technology*, Vol. 11, 1977, pp. 359-366.
[24] Zepp, R. G. and Schlotzhauer, P. F., "Photoreactivity of Selected Aromatic Hydrocarbons in Water," *Polynuclear Aromatic Hydrocarbons*, P. W. Jones and P. Leber, Eds., Ann Arbor Sciences, Ann Arbor, MI, 1979, p. 892.
[25] Bartell, S. M., Landrum, P. F., Giesy, J. P., and Leversee, G. J. in *Energy and Ecological Modelling*, W. J. Mitsch, R. W. Bosserman, and J. M. Klopatek, Eds, Elsevier, New York, pp. 133-150.
[26] O'Neill, R. V., Gardner, R. H., Barnthouse, L. W., Suter, G. W., Hildebrand, S. G., and Gehrs, C. W., "Ecosystem Risk Analysis: A New Methodology," *Environmental Toxicology and Chemistry*, Vol. 1, 1982, pp. 167-177.
[27] O'Neill, R. V., Bartell, S. M., and Gardner, R. H., "Patterns of Toxicological Effects in Ecosystems: A Modeling Study," *Environmental Toxicology and Chemistry*, Vol. 2, 1983, pp. 451-461.

R. Anne Jones[1] and G. Fred Lee[1]

Use of Vollenweider-OECD Modeling to Evaluate Aquatic Ecosystem Functioning

REFERENCE: Jones, R. A. and Lee, G. F., **"Use of Vollenweider-OECD Modeling to Evaluate Aquatic Ecosystem Functioning,"** *Functional Testing of Aquatic Biota for Estimating Hazards of Chemicals, ASTM STP 988,* J. Cairns, Jr., and J. R. Pratt, Eds., American Society for Testing and Materials, Philadelphia, 1988, pp. 17–27.

ABSTRACT: The Vollenweider–Organization for Economic Cooperation and Development (OECD) models are statistical relationships that describe the functioning of aquatic ecosystems in the use of aquatic plant nutrients (nitrogen and phosphorus) for the production of planktonic algae. These models were developed on the basis of data from several hundred bodies of water throughout the world; those discussed herein define the relationships between normalized phosphorus loads to lakes and reservoirs and the planktonic algal chlorophyll, algal productivity, algal-related water clarity, hypolimnetic oxygen depletion rate, and fish yield. Guidance is provided on the use of these relationships for assessing the overall functioning of an aquatic ecosystem.

KEY WORDS: hazard evaluation, ecosystem functioning, ecosystem modeling, phosphorus, algae, chlorophyll

One approach to the evaluation of whether an aquatic ecosystem is functioning as it should or the functioning has been altered, is to compare the functioning of the ecosystem of concern to some norm. In making this comparison, several issues must be addressed: what does "functioning of an ecosystem" mean; what parameters are used to assess this functioning; what is a "norm" for the particular type of system; what is the expected range of variability among "normally" functioning ecosystems of this type; and how far from the norm or normal range does the functioning of an ecosystem have to be before it is judged to be different enough from the norm to be of concern? In the case of ecosystem evaluation associated with water quality management, is being that different from the norm something of sufficient concern to society to justify contaminant control; can the same level of deviation from the norm perhaps occur through natural events unrelated to contaminant input?

There are various ways in which ecosystems and their functioning have been evaluated. For example, these evaluations have often been tied to the numbers and diversity of organisms included in a survey of a system. Other approaches focus on conducting specific laboratory tests, such as enzyme production or activity, to assess how well a particular part of a component of the system is operating. Other approaches attempt to develop, through laboratory microcosms, appropriate replicates of a full-scale ecosystem for study. A number of papers describing these types of approaches are presented in this volume.

The focus of this paper is on the application of the Vollenweider–Organization for Economic Cooperation and Development (OECD) modeling approach to the assessment of overall aquatic ecosystem functioning. These models describe, in a normalized statistical modeling framework,

[1]Associate professor and professor, respectively, Department of Civil and Environmental Engineering, New Jersey Institute of Technology, Newark, NJ 07040.

ecosystem functioning as described by the utilization of phosphorus load in the production of planktonic algal chlorophyll, primary productivity, water clarity as controlled by primary production, hypolimnetic oxygen depletion rate, and overall fish yield. The basis for these empirical models is the functioning of several hundred ecosystems; they serve as a reasonable description of a "norm" and range of expected ("normal") conditions.

Vollenweider-OECD Models

During the mid-1970s the Organization for Economic Cooperation and Development (OECD) sponsored a five-year study of the relationships between nutrient loading and eutrophication-related water quality response in bodies of water in the United States, Canada, Australia, Japan, and 14 countries in Western Europe. Vollenweider, one of the principal individuals directing this OECD study and the overall study coordinator, had already begun to investigate, quantify, and mathematically describe the factors influencing how planktonic algae in a water body utilized the nutrient input, based on a group of about 20 lakes, primarily European [1–3]. He found that the water-body mean depth, surface area, and hydraulic residence time play critical roles in the amount of planktonic algal biomass that develops within a water body for a given phosphorus load to that water body. On the basis of the data on the group of European water bodies, he defined a statistical regression between the epilimnetic planktonic algal chlorophyll concentration and the phosphorus load normalized by mean depth, hydraulic residence time, and surface area. The phosphorus loading is normalized in accordance with the following formulation

$$\frac{L(P)/q_s}{1 + \sqrt{\tau_w}} \qquad (1)$$

where

$L(P)$ = the areal annual total phosphorus load, mg P/m^2/year,
q_s = the mean depth, m, divided by the hydraulic residence time, year, and
τ_w = the hydraulic residence time, year.

Inherent in the mathematical/theoretical formulation of the normalization of the phosphorus load are the interactions of nutrients with the sediments and particulate matter in the water body, the availability of the phosphorus input (most of the water bodies included had on the order of 75% of the input phosphorus load in available forms), and the grazing of the phytoplankton. The normalized phosphorus loading term developed is theoretically equivalent to the average in-lake phosphorus concentration; based on data available on U.S. and international OECD water bodies, the in-lake phosphorus concentration can be mathematically described as a function of normalized phosphorus loading. Thus, the model simply mathematically describes the relationship between the nutrients available and the phytoplankton that develop, that is, algal stoichiometry. Similar stoichiometric relationships are reflected in the correlations between the phytoplankton biomass and the higher trophic levels described further on in this paper.

With the data for the OECD water bodies in the United States [4] and the international OECD water bodies [5], and the results of studies conducted since by the authors and their associates [6, 7], the basis for Vollenweider's empirical relationship has been expanded to more than 500 water bodies throughout the world (Fig. 1). These water bodies include lakes, reservoirs, and an estuarine system. They include shallow and deep, large and small, temperate and antarctic, and ultraoligotrophic and eutrophic water bodies, as well as those in between. These water bodies would, in general, be considered to be unaffected by toxic chemicals that would cause the overall aspects of their ecosystem functioning to be adversely affected.

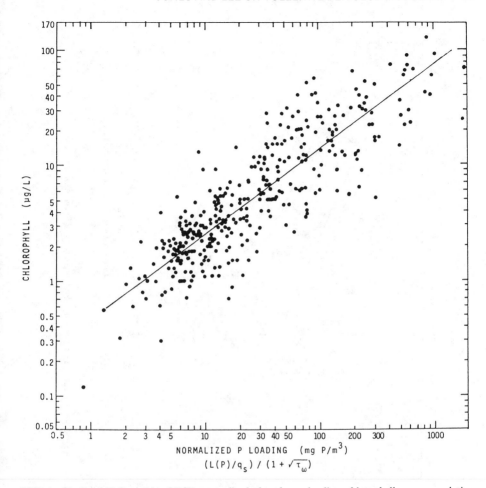

FIG. 1—*Updated Vollenweider-OECD normalized phosphorus loading–chlorophyll response relationship* [7] *for bodies of water throughout the world.*

With their definition of the empirical relationship between normalized phosphorus loading and planktonic algal chlorophyll for the OECD water bodies in the United States, Rast and Lee [4] expanded the approach and developed analogous relationships between the normalized phosphorus load and the Secchi depth (for water bodies with moderate or little inorganic turbidity or color) and also between the normalized phosphorus load and the hypolimnetic oxygen depletion rate. Jones and Lee [6] updated these statistical relationships to include data from other water bodies on which data were available; Fig. 2 shows these relationships.

In addition to the relationships shown in Figs. 1 and 2, work was being conducted independently to relate the planktonic algal biomass to the overall yield of fish in water bodies. Based on information from the literature [8,9] and the normalized phosphorus load-response relationships in Figs. 1 and 2, Lee and Jones [10] found and described the statistical relationship shown in Fig. 3 between the normalized phosphorus load and the overall fish yield. While the data base for this relationship is smaller than those for the other models, the trend found is as would be expected: the greater the phosphorus loading, the greater the chlorophyll-algal biomass, and the greater the overall fish yield. Lee and Jones [10] and Jones and Lee [7] discussed the develop-

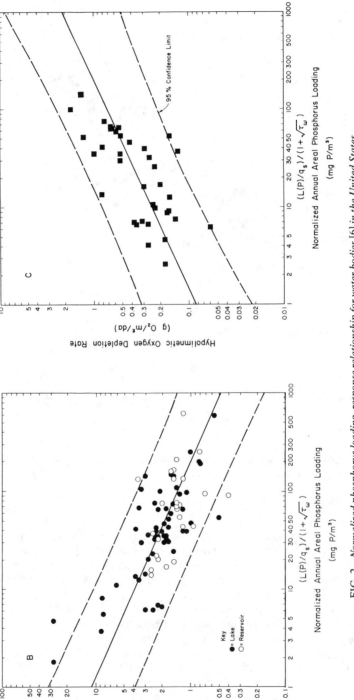

FIG. 2—*Normalized phosphorus loading–response relationship for water bodies [6] in the United States.*

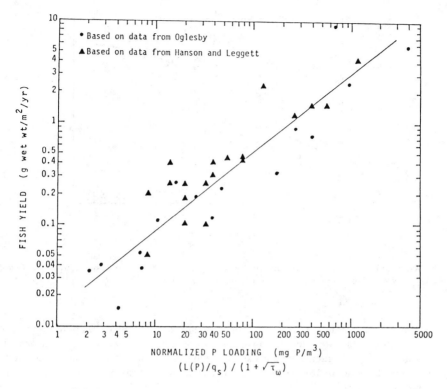

FIG. 3—*Normalized phosphorus loading–fish yield relationship* [10].

ment and interpretation of this relationship. Vollenweider[2] [5] described a relationship between the normalized phosphorus load and the primary productivity of the OECD water bodies. This relationship is presented in Fig. 4. Vollenweider[2] has indicated that a straight line could equally well be used to describe the function shown.

One of the most important aspects of the Vollenweider-OECD modeling approach is the demonstration and verification of the predictive capability of the relationships. Rast, Jones, and Lee [11] undertook a study to evaluate the data from studies of nutrient load and planktonic-algal-related response for water bodies that had undergone phosphorus load reductions, in order to evaluate the predictive capability of the models. They examined and plotted the data that described the load-response couplings before and after substantial phosphorus loading alterations had occurred in about a dozen water bodies for which appropriate data were available. The results of this evaluation for the chlorophyll response are shown in Fig. 5, which demonstrates that, in general, with alteration in their normalized phosphorus loadings, water-bodies track parallel to the line of best fit through the body of data upon which the relationship was developed (Fig. 1). For essentially all of the water bodies evaluated, the predicted response values and the measured response values after the loading change were within a factor of 1.5 of each other. Rast et al. [11] and Jones and Lee [7] provide a discussion of these results and their interpretation for water quality management.

In the use of the Vollenweider-OECD modeling approach for evaluating overall ecosystem functioning, it is important to understand the variability of the data about the lines of best fit

[2]Vollenweider, R. A., Canada Centre for Inland Waters, Burlington, Ontario, Canada, personal communication, 1986.

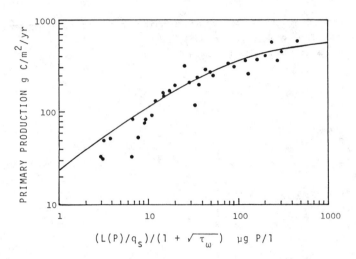

FIG. 4—*Relationship between the normalized phosphorus load and primary productivity. The figure is from Vollenweider, Canada Centre for Inland Waters, Burlington, Ontario, 1979.*

defining the individual relationships. Limitations on and concerns about the applicability of the modeling approach for water quality management have been addressed by Rast et al. [11] and by Jones and Lee [7]. The family of points about the line of best fit (noted by confidence interval lines on some of the figures) shows how the water bodies' load-response couplings are related and also how they vary.

Use of Models for Evaluation of Overall Aquatic Ecosystem Functioning

The Vollenweider-OECD load-response models can be used to address and integrate several levels of overall "ecosystem functioning" by indicating whether the normalized phosphorus load–response coupling for a water body falls within the family of couplings for the several hundred water bodies unaffected by toxicant input, upon which the models were developed. They can indicate whether the phosphorus loading supports the amount of planktonic algal biomass and productivity expected based on the morphologic and hydrologic characteristics of the water body, whether the rate of bacterial decomposition in the hypolimnion (oxygen depletion rate) is as expected based on the amount of algal growth, and whether the fish yield is as expected based on the amount of algal biomass in the water body. They can also indicate whether the water body responds to phosphorus load alterations as expected [11].

In order to apply this approach, the phosphorus loading must be determined. This can be done, as described by Rast and Lee [12], based on land use and phosphorus export coefficients or on measurements of water inflow and phosphorus concentrations from the major sources to the water body. (Internal cycling is accounted for in the normalizing.) The mean depth, hydraulic residence time, and surface area must also be estimated. The mean depth is equivalent to the volume of the water body divided by the surface area; the hydraulic residence time can be determined by dividing the volume by the annual water inflow; and the normalized phosphorus loading is computed as indicated in Eq 1.

An estimate of the overall water-body ecosystem response must also be made. The best measurement for the purposes of this modeling is usually the average summer epilimnetic planktonic algal chlorophyll. For water bodies that do not have excessive amounts of inorganic turbidity or color, the average summer Secchi depth is also a suitable response parameter, as it is directly related to the planktonic algal chlorophyll. The models themselves can be used to assess

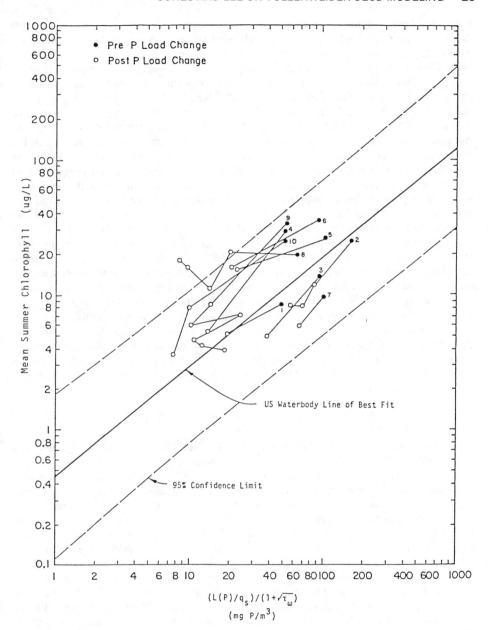

FIG. 5—*Normalized phosphorus loading–chlorophyll response couplings before and after phosphorus load changes* [11].

whether the inorganic turbidity or color is significantly limiting phytoplankton production in a water body [7]. Primary productivity and fish yield can also be utilized as response parameters, but the data bases for the models for these parameters are not as well developed at this time as those for chlorophyll and Secchi depth.

The normalized phosphorus loading and response couplings are then plotted on the appropriate model. If the coupling coordinates are within the family of points that comprise the model, it may be concluded that the aspect of the ecosystem considered is functioning as may be expected based on the norm established by the model. If the coordinates are outside the family of points, the data inputs should be reexamined: First, the reliability of all input data should be examined. Then, each aspect of data input (phosphorus load, morphology, hydrology, and response), should be evaluated to determine whether a factor that would be expected to skew the relationship was overlooked in applying the model. Jones and Lee [6, 7] and Rast et al. [11] discuss the conditions under which the models are not applicable or may need to have input data modified before application. As discussed in those sources, the points of focus should include the following: If the water body has a hydraulic residence time of less than two weeks, if it has a high degree of inorganic turbidity or color (a Secchi depth of less than about 0.5 m during non-algal-bloom periods), or if primary production is significantly manifested as aquatic macrophytes or nonplanktonic algae, the models should not be applied directly. If the water body has significant arms or bay areas or is long and thin in shape, with the result that the nutrients added do not rapidly mix throughout the water body, then it should be sectioned and the sections should be evaluated separately for their load-response relationships. If there has been a major perturbation in the phosphorus loading within the recent past (typically during the two to three years prior to the load-response assessment), the water body may not have come to equilibrium with its new loading. In that case, the water body response could be reflecting past loading conditions and not the loading that was measured or estimated at the time of the response assessment. It has been found [11, 13] that a new equilibrium is established after a time period equivalent to about three times the phosphorus residence time of the water body.

If the reevaluation does not reveal conditions that would render the model inappropriate, and the data appear reliable, it may be concluded that something is causing the water body ecosystem to be functioning differently from the approximately 500 water bodies on which the models were formulated. This would suggest that there may have been an impact on the functioning of the aspect of the ecosystem assessed.

Application of Models

One useful application of the Vollenweider-OECD models in overall ecosystem functioning assessment is in the management of toxics in aquatic systems. This is an area of particular importance to the U.S. Environmental Protection Agency (EPA) and state water pollution control agencies, which are beginning to implement control programs for toxics in industrial and domestic wastewaters [14]. One of the most important questions that has to be addressed in the cost-effective implementation of these programs is the potential benefits that can be derived from the control of toxicity in a wastewater effluent in terms of overall ecosystem functioning as it relates to designated beneficial uses of the receiving waters. If the toxicity is restricted to an area near the point of discharge, that is, the mixing zone, it may be of little or no significance to the overall ecosystem. However, if the impact is system-wide and directed toward system components of concern to the public, such as the numbers of catchable desirable fish, then much greater emphasis should be given to developing control programs. While ecosystem functioning evaluation approaches that focus on measurement of one aspect of the ecosystem may indicate that an impact has occurred, especially if evaluated near the point of discharge of a toxic effluent, this impact may be of little or no consequence to the functioning of the *overall* ecosystem. The Vollenweider-OECD modeling approach allows evaluation of the overall functioning of a number of integrated aspects of the ecosystem. Lee and Jones [15] have discussed the problems of implementing the U.S. EPA [14] toxics control program as it relates to beneficial uses of receiving waters.

An example of how the Vollenweider-OECD models can be used for overall ecosystem functioning evaluation is seen in work the authors of this paper are doing in the New York/New Jersey coastal waters. Preliminary evaluations of the phosphorus load and chlorophyll response for the Hudson/Raritan estuary in New York and New Jesery show that less algal biomass is being produced than would be expected based on the normalized phosphorus loading. The authors suspect that this situation may be related to the presence of toxics in these waters. However, what will happen to eutrophication-related water quality as the concentrations of toxics are reduced is of concern to the recreational, commercial, and residential developments in that area. It is believed that as the toxics are reduced, as they will be with increased control, the problems with excessive fertilization will increase; the algae will be able to grow to the levels expected based on the loading of nutrients and the other characteristics of the ecosystem. Therefore, unless phosphorus control programs are initiated along with toxics control programs, the development projects and existing uses, which are tied to the aesthetic character of these waters, may be adversely impacted by toxics control programs. The Vollenweider-OECD modeling approach can be used to determine how much phosphorus control is needed to achieve a given level of planktonic algal production in a particular water [7]. In addition to the concern for eutrophication problems resulting from the toxics control programs, the model also indicates that the toxics present at this time may be in sufficient quantities to limit the ability of the system to support fish, because of decreased amounts of planktonic algae.

It has also been found that, in laboratory toxicity tests, water extracts from sediments of the Hudson/Raritan estuarine system have substantial toxicity to algae and varied (generally low to moderate) toxicity to higher level organisms. The approach discussed herein could be used to assess the overall impact on ecosystem functioning associated with this source of contamination, and to help assess the cost-effectiveness of techniques that may be considered for remediation and future handling of the contaminated sediments.

Sensitivity of Relationships

In assessing the potential utility of the Vollenweider-OECD modeling approach (or any approach) for overall ecosystem functioning evaluation in a particular water, it is important to consider the sensitivity of a potential management decision to the results of the evaluation. Ecosystem functioning and the characteristics used to assess ecosystem functioning may be highly variable within any one system, and may vary considerably with the season, year, or other naturally occurring conditions. The scatter of points about the lines of best fit in the models illustrated attest to the natural variability in ecosystem functioning and to the variability that results from constraints in our ability to define the relationships precisely. It should be noted that a considerable part of the variability about the lines of best fit relates to the accuracy with which measurements of chlorophyll and phosphorus loads can be made.[3] Nonetheless, since chlorophyll and phosphorus determinations are among the more reliable measurements typically made, the reliability of this approach is likely to be at least as good as the reliability of other, less well established methods of assessment of ecosystem functioning.

Variability may also be associated with the differences in manifestations of primary production in any given water body. The response parameters used in these models are associated with planktonic algal growth. The extent to which unusual proportions of the primary production are in nonplanktonic algal forms may contribute to some of the scatter. These characteristics, however, may vary over an annual cycle or from year to year. Even with whatever variation exists in the accuracy of chlorophyll and phosphorus load measurements, and with variations in the conditions of the water bodies, the predictive capability of this approach has been clearly demonstrated [11], with the predicted values of planktonic algal chlorophyll found to be within 50% of the measured values in real-world situations. This type of verification is not available for most types of environmental quality models when applied to systems other than those for which they

[3]Vollenweider, R. A., Canada Centre for Inland Waters, Burlington, Ontario, Canada, personal communication, 1983.

were developed and tuned. It illustrates the solidity of the models and their applicability despite normally expected variations in planktonic algal chlorophyll and other measurements.

As an example of the sensitivity of the model, examination of Fig. 1 shows that a water body with a normalized phosphorus loading of 100 mg phosphorus/m^3 would be expected to have an average chlorophyll concentration of between 4 and 60 $\mu g/L$. For a water body on which minimal data were available, if the average chlorophyll concentration were substantially out of that range, one could consider the system to behave differently from other systems evaluated. This may seem to be a rather insensitive indicator, with more than an order of magnitude in a "normal" characteristic range. However, several things must be considered before that assessment can be deemed to be appropriate. First, based on a considerable body of data, the range of the "norm" for the variety and number of water bodies included is consistent throughout the three orders of magnitude of normalized phosphorus loading for which data are available. It thus appears that the family of points describing the norm is, indeed, an appropriate assessment of the degree of variability that can be expected in the chlorophyll response—or overall ecosystem functioning—to normalized phosphorus loading. The data base for the models is extensive and expansive and, for most of the water bodies included, is based on a substantial monitoring program. Second, the sensitivity of this overall ecosystem functioning evaluation approach must be viewed in relation to the sensitivity of other ecosystem functioning evaluation approaches available when extrapolated to overall ecosystem functioning. The authors are not aware of any other ecosystem functioning modeling approach that has been evaluated for the variety and number of water bodies that the Vollenweider-OECD modeling approach has.

Third, and most important, if a monitoring program is established on the water body in question, in which load and response are measured over a several-year period, the location of the couplings for the water body from year to year can be more closely defined. As discussed by Jones and Lee [7], having this data base can reduce the range of the norm for the particular water body in question considerably. However, the behavior of the water body in its phosphorus load-response relationship in the model, the external perturbations, and the recovery time for perturbations must be well understood before load-response couplings that fall within the family of points shown in Fig. 1 are judged to represent substantial effects on overall ecosystem functioning.

Summary and Conclusions

The Vollenweider-OECD models provide a synthesis indication of overall ecosystem functioning. They can describe whether as many algae grow as would be expected based on the normalized phosphorus loading or whether something is causing fewer algae to grow. They also describe whether as much fish biomass is supported by the algal biomass measured or expected as would be expected. In addition, they provide insight into the effects of perturbations on the functioning of the ecosystems assessed, such as nutrient load reductions or increases, or changes in the morphological or hydrological characteristics of the water body. These relationships have demonstrated capability to predict the impact of such perturbations on the aspects of the overall ecosystems evaluated. The models are based on the actual load-response relationships found in a wide variety of aquatic ecosystems throughout the world.

While some ecosystem function tests are of limited scope in that they only assess a small portion or aspect of the system, or rely on a laboratory model ecosystem, the Vollenweider-OECD modeling approach can assess a variety of functions and, indeed, integrates them into an overall system assessment. In the use of any ecosystem functioning test or evaluation approach for environmental management purposes, this integration needs to be made. Only in the context of how a perturbation has affected the overall integrity of an ecosystem—a much broader context than one aspect of the system—can effective management programs be developed. This is the focus of the Vollenweider-OECD models.

Acknowledgment

The contributions of R. Vollenweider toward developing the nutrient load-response models are acknowledged. Support for preparation of this paper was provided in part by the Department of Civil and Environmental Engineering at the New Jersey Institute of Technology.

References

[1] Vollenweider, R. A., "Scientific Fundamentals of the Eutrophication of Lakes and Flowing Waters with Particular Reference to Nitrogen and Phosphorus as Factors in Eutrophication," Technical Report DA 5/SCI/68.27.250, Organization for Economic Cooperation and Development, Paris, France, 1968.

[2] Vollenweider, R. A., "Input-Output Models with Special Reference to the Phosphorus Loading Concept in Limnology," *Schweiz. A. Hydrol.*, Vol. 37, 1975, pp. 53–84.

[3] Vollenweider, R. A., "Advances in Defining Critical Loading Levels for Phosphorus in Lake Eutrophication," *Mem. Ist. ital. Idrobio.*, Vol. 33, 1976, pp. 53–83.

[4] Rast, W. and Lee, G. F., "Summary Analysis of the North American (U.S. Portion) OECD Eutrophication Project: Nutrient Loading–Lake Response Relationships and Trophic State Indices," EPA-600/3-78-008, U.S. Environmental Protection Agency, Washington, DC, 1978.

[5] "Eutrophication of Waters—Monitoring, Assessment, and Control," Organization for Economic Cooperation and Development, Paris, France, 1982.

[6] Jones, R. A. and Lee, G. F., "Recent Advances in Assessing the Impact of Phosphorus Loads on Eutrophication-Related Water Quality," *Journal of Water Research*, Vol. 16, 1982, pp. 503–515.

[7] Jones, R. A. and Lee, G. F., "Eutrophication Modeling for Water Quality Management: An Update of the Vollenweider-OECD Model," *World Health Organization Water Quality Bulletin*, Vol. 11, No. 2, 1986, pp. 67–74, 118.

[8] Oglesby, R. T., "Relationships of Fish Yield to Lake Phytoplankton Standing Crop, Production, and Morphoedaphic Factors," *Journal of the Fisheries Research Board of Canada*, Vol. 34, 1977, pp. 2271–2279.

[9] Hanson, J. and Leggett, W., "Empirical Prediction of Fish Biomass and Yield," *Canadian Journal of Fisheries and Aquatic Sciences*, Vol. 39, 1982, pp. 257–263.

[10] Lee, G. F. and Jones, R. A., "Impact of Eutrophication on Fisheries," CRC Press, in press.

[11] Rast, W., Jones, R. A., and Lee, G. F., "Predictive Capability of U.S. OECD Phosphorus Loading-Eutrophication Response Models," *Journal of the Water Pollution Control Federation*, Vol. 55, 1983, pp. 990–1003.

[12] Rast, W. and Lee, G. F., "Nutrient Loading Estimates for Lakes," *Journal of Environmental Engineering Division of the American Society of Civil Engineers*, Vol. 109, 1983, pp. 502–517.

[13] Sonzogni, W. C., Uttormark, P. C., and Lee, G. F., "A Phosphorus Residence Time Model: Theory and Application," *Water Research*, Vol. 10, 1976, pp. 429–435.

[14] "Technical Support Document for Water Quality-Based Toxics Control," EPA-440/4-85-032, U.S. Environmental Protection Agency, Washington, DC, 1985.

[15] Lee, G. F. and Jones, R. A., "Assessment of the Degree of Treatment Required for Toxic Wastewater Effluents," *Proceedings*, International Conference on Innovative Biological Treatment of Toxic Wastewaters, U.S. Army Construction Engineering Research Laboratory, Champlain, IL, 1987, pp. 652–677.

Dick de Zwart[1] and Hennie Langstraat[1]

Chemically Induced Community Responses in a Compartmentalized Microcosm Assessed by Multidimensional State Space Transitions

REFERENCE: de Zwart, D. and Langstraat, H., **"Chemically Induced Community Responses in a Compartmentalized Microcosm Assessed by Multidimensional State Space Transitions,"** *Functional Testing of Aquatic Biota for Estimating Hazards of Chemicals, ASTM STP 988,* J. Cairns, Jr., and J. R. Pratt, Eds., American Society for Testing and Materials, Philadelphia, 1988, pp. 28-40.

ABSTRACT: A test has been developed to assess the chemically induced disturbance produced by interactions between functionally different groups of organisms. Algae, daphnids, polyps, and decomposer organisms—representing adjacent trophic levels—were housed in separate containers. Water containing algae and nutrients was circulated through the compartments. In some experiments, according to a realistic scheme based on empirically derived knowledge of the needs of the individual populations, an interactive relationship was maintained between standing stock parameters and the circulation rate and between predation on daphnids and feeding by polyps. Since, for each compartment, input and output fluxes were experimentally determined, the rates of processes could be calculated. One of the major drawbacks of the small volume (14 L) of the microcosm was the necessity for nondestructive or micro methods of analysis.

In the experiments presented in this paper, only the standing stock parameters for the algae, daphnids, and polyps, and occasionally the pH, were measured with enough adequacy to construct a multivariate "normal operating range" (95% confidence interval) out of the rather long time series of preexpositional data. This "normal operating range" was used as a reference state for the changes after perturbation.

Perturbation of the model ecosystem with a toxic chemical (in this case, 2,4-dichloroaniline) was followed by a gradual displacement outside the normal operating range. All four systems tested showed a temporary or permanent shift in the three-dimensional state space outside the normal operating confinement within ten days after the addition of 0.010 mg L^{-1} of the toxicant, whereas the lowest "no-observed-toxic-effect level," determined by semichronic tests on the susceptibility, individually, of populations of green algae, daphnids, and polyps, was slightly higher. (For *Daphnia magna,* this level was 0.032 mg L^{-1}.)

KEY WORDS: hazard evaluation, ecotoxicology, model ecosystem, species interactions, multivariate statistics

In traditional aquatic toxicology, individual populations of test species are exposed to chemical compounds. Stresses other than the one to be tested are generally omitted, and the test medium is renewed frequently to exclude the influence of possible transformation products. For this reason standard toxicity tests can only provide information on the relative toxicity of chemi-

[1]Research ecotoxicologist and ecologist, respectively, Laboratory for Ecotoxicology, Environmental Chemistry, and Drinking Water, National Institute of Public Health and Environmental Hygiene, 3720 BA Bilthoven, The Netherlands.

cals. However, in the natural environment the stress resulting from the introduction of man-made chemicals is added to such stresses as food shortage, competition, chemical composition, and temperature. The tendency of all living creatures to exploit their own niche to the maximum implies that a balanced ecosystem is regulated by stresses resulting from biotic, chemical, and physical interactions. What is really of interest is the effect of perturbation on the functioning of an ecosystem as a whole. Special problems arise when investigating the effects of environmental perturbations on real natural ecosystems. Because of the uncontrollable variability in environmental conditions, it is questionable whether subtle indications, generally preceding gross disruptions, can be adequately detected. Severe experimental perturbations, needed to invoke acutely detectable effects, are thought to be unacceptable, owing to the risk of irreversible damage to the system studied. The uniqueness of each ecosystem presents another problem in assessing the effects of pollution. As it is totally impossible to assess the effects of all kinds of pollution on all ecosystems, the only feasible alternative is to assess the effects on general ecosystem properties, which, when dealing with subtle perturbations, are best studied under controlled conditions. The solution is to bring part of a system to the laboratory, retaining as much of its essential complexity as possible, including especially the producer, consumer, regenerator, and circulating phases [1]. According to Giesy and Odum [2], this kind of model ecosystem or microcosm has to be regarded as "artificially bounded subsets of naturally occurring environments which are replicable . . . [and] should exhibit system-level properties." Bringing parts of natural systems to the laboratory inevitably introduces artifacts as a result of isolation and scaling. Only effects on those properties and interactions shared by the experimental model with real-world ecosystems allow direct extrapolation to the field. A microcosm inevitably will be the product of a compromise between the irreconcilable desires for replicability (and therefore simplicity) and for resemblance to natural systems (and therefore complexity).

The rationale for using microcosms in environmental impact prediction is presented in Fig. 1. Site-specific evaluation and impact prediction require mathematical modeling based on ecosystem generalizations, microcosm validation, and criteria on local vulnerability [3], expressed in the following terms:

(a) the tolerance limits in the original situation (inertia),
(b) the ability to recover from damage (elasticity),
(c) the ability to return to the original situation after a perturbation (resiliency).

In addition to evaluation of the experimental design, the main objective of the present study was the application and modification of a statistical method for detecting and scaling the deviation from normal functioning (strain) of chemically stressed microcosms. Such a method is fairly indispensable, because, even under strictly controlled conditions, it is very difficult to distinguish strain from the seemingly random variability of interacting parameters. The method produces a time series of a single score for system strain, which, in relation to the amount of chemicals added, can be used to quantify the inertia, elasticity, and resilience of the microcosm.

Methods

The Microcosm and Operating Conditions

The idea behind the type of microcosm used was developed by Ringelberg and Kersting [4]. Compared with a natural system, a relatively simple aquarium lacks the spatial heterogeneity needed to prevent disorganized overconsumption. Therefore, the microcosm [or micro-ecosystem (MES)] used consisted of four separate subsystems which, ideally, modeled several different trophic levels: autotrophs (phytoplankton species), herbivores (*Daphnia magna* Straus), carnivores (*Hydra oligactis* Pallas), and decomposers (fungi and bacteria). Each subsystem was housed in a glass container (Sovirel, Corning Ltd., Staffordshire, United Kingdom), which was

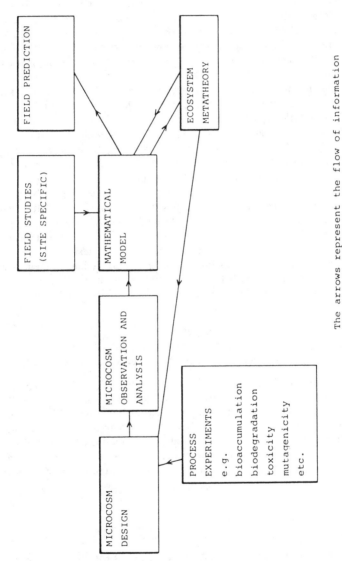

FIG. 1—*Rationale for the use of microcosms in environmental impact prediction.*

The arrows represent the flow of information required to perform the boxed tasks.

interconnected with the other subsystems by Teflon tubing (Fig. 2). All the subsystems were placed in a water bath providing a temperature of $17 \pm 1°C$. The algal compartment was filled with 9 L of 0.45-μm-membrane (Sartorius-Membranfilter GmbH, Göttingen, Federal Republic of Germany)-filtered water from Lake Reeuwijk, The Netherlands. The herbivore and carnivore compartments both contained 2 L of filtered water, whereas the decomposer compartment was filled with 1 L of acid-washed sand and 1 L of filtered water. From underneath the sand bed a peristaltic pump (Microperpex, LKB-Produkter BA, Bromma, Sweden) fed the water containing mineralized nutrients to the autotrophic compartment at a predetermined rate. Since the four compartments were sealed, displacement of water invoked a circulation flow. Water containing algae was fed to the herbivore compartment, and water containing algae and other suspended materials flowed through the carnivore compartment back to the water layer on top of the sand bed, where settling occurred and mineralization took place. The presence of microorganisms responsible for mineralization was guaranteed by seeding the decomposer compartment with 10 mL of bottom sediment from Lake Reeuwijk. A few individual daphnids and polyps were furnished once from laboratory cultures, shortly after the system was filled. Algae were inoculated into the first compartment once every 14 days using deep-frozen "algal pills," which contained 1 mL of a viable mixture of different algae from Lake Reeuwijk (green and blue-green algae, and diatoms). The method of reinoculation with the diverse mixture of algae was used to prevent rapid shifts in species composition caused by accidental infections with species that, if present earlier, would already be dominating. In view of the diversity of the introduced algae, a shift in species composition could probably be ascribed to chemical perturbation instead of infection.

In the autotrophic subsystem, the algae grew under the influence of nutrients and the light of six fluorescent lamps (20-W True-Light preheat power twist tubes, Duro-Lite International, Fairlawn, New Jersey). During 12 h per day the light intensity ranged from 150 to 250 μeinsteins m^{-2} s^{-1} inside the autotrophic compartment. In the herbivore and carnivore compartments, the light intensity was about 20 μeinsteins m^{-2} s^{-1}, whereas the decomposer compartment was constantly in the dark.

A predetermined part of the standing stock in the herbivore compartment was manually fed twice a week to the hydra population. The compartmentalization of the microcosm required that all material be involved in circulating through the four stages. To ensure the necessary dominance of planktonic algae, therefore, conditions were made unfavorable for wall-growing

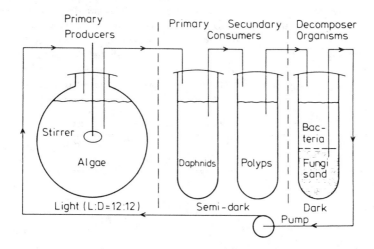

FIG. 2—*Schematic representation of the microcosm.*

algae by stirring the autotrophic compartment vigorously (Vibro-Mischer E-1, Chemap AG, Männedorf, Switzerland). In the other compartments, all debris excluded from normal circulation was transferred manually to the decomposer compartment twice a week.

When the circulation rate and the percentage of predation on the herbivores were held constant, there was no influence from the population size of the herbivores on the algae or of the carnivores on the daphnids. This is not a very realistic property of the model. In a noncompartmentalized natural system, an increasing population of consumers will put more predation stress on the population of their food organisms. Therefore, in two out of three experiments performed in 1983 (Microcosms MES83-1 and -3) the circulation rate was controlled by the size of the herbivore population, and the predation rate on the herbivores was controlled by the size of the carnivore population. The dependencies outlined in Table 1 are based on empirically derived knowledge of the performance and needs of the individual populations [5,6]. The dilution rate of the algae and the supply of nutrients had to be within a range that could maintain a biovolume of 0.01 mm^3 mL^{-1}. (The density of the algae in Lake Reeuwijk was 0.005 to 0.015 mm^3 mL^{-1}.) At the maximum circulation rate of 60 mL h^{-1}, the retention time in the algal compartment was approximately one week. Determination of the maintenance energy required by *Daphnia magna* [5] has revealed that an algal supply of 60 × 0.01 mm^3 h^{-1} is adequate to sustain a population of about 600 mature daphnids in the 2-L herbivore compartment (which is realistic with respect to the carrying capacity of 500 to 1000 individuals per litre, both young and adult, often observed in laboratory cultures).

Microcosm MES83-2 was operated at a constant flow rate of 60 mL h^{-1} and 10% predation on the herbivore population per week. From the experiments performed in 1985, only Microcosm MES85-2 was used to study effects, whereas the MES85-1 and MES85-3 systems were devoted to a fate study. Up until 156 days after the start of the experiment, MES85-2 was operated at a flow rate of 60 mL h^{-1}, which was then reduced to a constant flow of 30 mL h^{-1} in order to elicit the effects of the manipulations of the flow rate just described more clearly. Predation on the herbivore population was kept constant at 10% per week.

In view of the limitations of time and the number of microcosms, together with the frequently observed dissimilar development of replicate microcosms of this kind [7,8], priority was given to investigation of the effects of experimental manipulations on the functioning of the microcosms above and beyond other attempts at creating replicates. Instead of using replicate microcosms, the preexpositional state of each system served as the control for that system. According to Vervelde and Ringelberg [9], this approach is only valid when this preexperimental state is a (dynamic) steady state. The effects of perturbations are expressed by the change in interactions and rates of processes once a new steady state is reached. Therefore, the systems were chemically stressed by small amounts of 2,4-dichloroaniline (DCA) after a period of relatively stable values of the parameters measured. From stocks of DCA in dimethylsulfoxide (DMSO), 1.4 mL of DMSO with appropriate concentrations of DCA was distributed over the compartments in proportion to their water volumes. With intervals of three to four weeks, subsequent doses of 10, 22, and 68 µg of toxicant per litre of microcosm water were administered, in an effort to deter-

TABLE 1—*Variable design characteristics of the model microcosm.*

Number of Daphnids		Flow Rate, mL h^{-1}	Number of Polyps		Predation of Daphnids, % daphnids week^{-1}
0 to 100	→	30	0 to 20	→	5
101 to 200	→	33	21 to 60	→	10
201 to 300	→	36	61 to 120	→	15
1000 and up	→	60	121 and up	→	20

mine the lowest dose that causes changes. In these experiments, DCA was chosen as the model test substance because considerable data already existed on the long-term susceptibility of the individual organisms incorporated in the microcosm.

Analytical Methods and Data Evaluation

To identify the performance of the populations involved, the number of daphnids and polyps and the biovolume of the algae were determined twice a week. The entire populations of the animal species were counted by eye, whereas the algae were volumetrically quantified by means of a Coulter counter combination (Model ZB plus C1000 Channelyzer plus Accucomp program, Coulter Electronics Ltd., Harpenden, United Kingdom) capable of producing computer-cumulated size-frequency distributions. The algal species composition; diurnal oxygen pulse, as a model for primary production in the autotrophic subsystem; nutrient cycling; and difference in pH between the outflow and inflow of the algal compartment (Δ pH), as a model for the metabolic activity of the heterotrophic part of the systems, were determined less frequently. Fruitless attempts were made to analyze the DCA concentration by means of spectrophotometric and gas-chromatographic methods. Since a sample volume of about 0.5 L could not be taken without disturbing the microcosm, the minimum detection limit available appeared to be about 100 μg L^{-1}.

However, for the construction of a normal operating range (NOR) and the quantification of strain, the biovolume of the algae and the numbers of daphnids and polyps proved to be the only parameters measured with enough adequacy in most cases. In Microcosm MES85-2 the pH values were continuously registered successfully enough to be included in the data evaluation. Evaluation of more parameters will give a better description of the microcosm state. Unfortunately, the polyps in this system did not develop into a stable population and had to be excluded from the evaluation. As a consequence, the NOR for the MES85-2 system has been constructed from the time series on the algae biovolume and the number of daphnids, in combination with the Δ pH values.

The NOR, which is based on the dependencies of the parameters included, is defined as the multidimensional 95% confidence interval of the trajectory produced by an undisturbed system in the orthogonal state space of those parameters [10]. In other words, 95% of the time an undisturbed system will show a combination of parameter values within the NOR. For three highly correlated parameters, the NOR takes the shape of a flattened ellipsoid. The inclusion of more state variables more strictly defines the state of the system but makes it impossible to visualize the NOR. The computational method, however, will be the same. The calculation of the NOR is based on the single-sample Hotelling $T^2_{0.95;p,N-p}$ statistic, resulting in a p-variate 95% confidence region based on the time series with N data points for each of the p state variables [11].

Perturbation of a system may induce a state vector outside this NOR. The location of this state vector with respect to the NOR is indicative of the strain the system suffers at this time. The euclidean distance A of the state vector and the center of the NOR (the mean vector of the state vectors found during stable unperturbed functioning) can be compared with the euclidean distance B of the center of the NOR and the intercept of the contours of the NOR and the line connecting the NOR center and the state vector [10]. Analogous to the Mahalanobis distance [11], the ratio of these two distances, A/B, can be used in the following classification rule: values from 0 to 1 indicate the state of the system within the NOR; values greater than 1 indicate a state different from the NOR. Following Kersting [10], we call this value normalized strain.

Results

In Fig. 3 the volume of algae and the number of daphnids and polyps in Microcosm MES83-1 are plotted against time. From Day 90 until the addition of DCA on Day 153, only a little

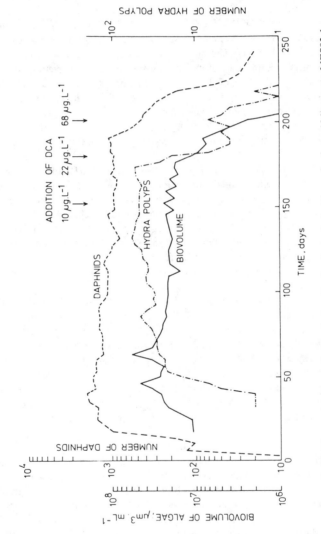

FIG. 3—*Time series of standing stock data for the algae, daphnids, and polyps of Microcosm MES83-1. The arrows indicate the moments of toxicant addition.*

variation was observed. After an exponential growth phase during the first days of the experiment, the growth of algae apparently stabilized as a result of either a stable supply of nutrients or a light limitation. There was an equilibrium between the net production of algae and the flow rate and between the net production of daphnids and their consumption by polyps. The values of the variables that were found in this period are supposed to represent the normal operating status of the microcosm. Because a similar pattern was found for Microcosms MES83-2 and MES83-3, the NOR of these microcosms could be based on observations during the same period. For the computation of NORs and normalized strain values, all 1983 data were transformed logarithmically to give the same weight to factorial increases and decreases of the standing stock parameters.

The normalized strain of Microcosm MES83-1 is plotted against time in Fig. 4. After Day 175, 23 days after the addition of the dose of 10 μg L^{-1} DCA, the normalized strain values were continuously greater than 1, indicating a permanent transition of the state of this microcosm outside its NOR. Two small temporary transitions outside the NOR occurred in the preceding period after the addition of DCA, but these transitions lasted only a few days. After Day 175 a nearly continuous increase of the strain was observed.

The normalized strain of Mircocosm MES83-2, plotted against time in Fig. 5, temporarily became greater than 1 after the addition of 10 μg L^{-1} DCA. Shortly after an extra dose of 22 μg L^{-1}, a transition outside the NOR of somewhat longer duration occurred. Ten days after this second addition of DCA, the normalized strain sharply increased, but returned to the NOR after a week. A fourth temporary transition preceded the permanent transition outside the normal operating range at Day 206, five days after the addition of an extra 68 μg L^{-1}. The ultimate strain was not as high as in the other two 1983 experiments.

A permanent transition of the state of Microcosm MES83-3 outside the normal operating range was already observed two days after the first addition of DCA (Fig. 6). As for MES83-1, a rapidly increasing normalized strain was found.

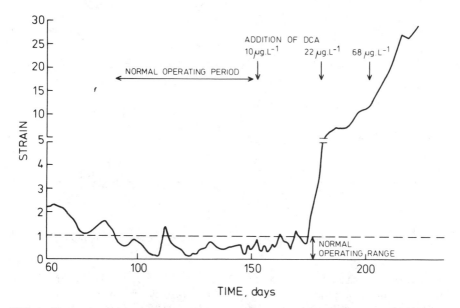

FIG. 4—*Time series of three-dimensional normalized strain in Microcosm MES83-1. The normal operating range is based on the standing stock data of Fig. 3 between Days 90 and 150 (the normal operating period). Normal operation is indicated by strain values between 0 and 1. The arrows indicate the moments of toxicant addition.*

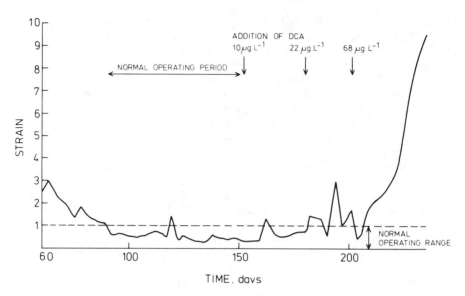

FIG. 5—*Time series of three-dimensional normalized strain in Microcosm MES83-2. The raw data on the standing stock parameters are not presented.*

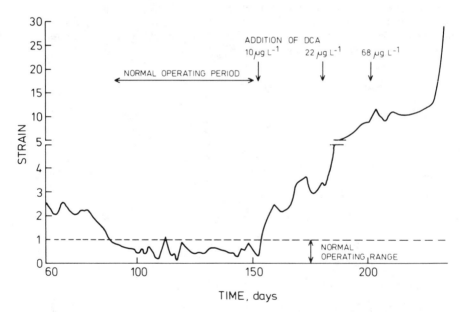

FIG. 6—*Time series of three-dimensional normalized strain in Microcosm MES83-3.*

The time series of the standing stock data of Microcosm MES85-2, presented in Fig. 7, reveal that the development of polyps was very unsuccessful. The carnivore compartment had to be reinoculated several times (Days 68, 156, 209, and 254). Figure 7 also contains data on the pH in the autotrophic compartment and the pH of the inflow into this compartment. For the calculation of the NOR and the strain, the availability of these adequately measured and interpretable data series made it possible to replace the log-transformed numbers of polyps by the non-

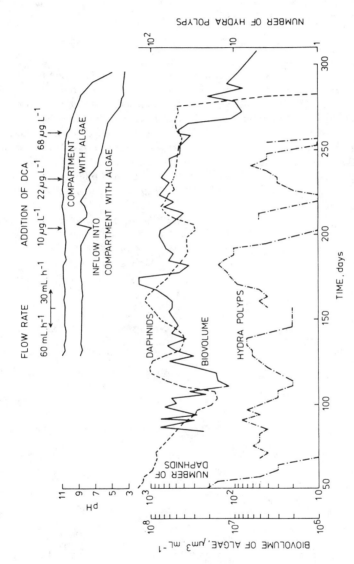

FIG. 7—*Standing stock data for the algae, daphnids, and polyps and data for the pH of the autotrophic compartment and its inflow for Microcosm MES85-2 against time. The moments of changing flow rate and the addition of toxicants are indicated.*

transformed absolute difference in both pH values. For Microcosm MES85-2, the NOR of the state variables was based on the observations between Day 156, when the flow rate was reduced to 30 mL h^{-1}, and Day 203, when the first dose of DCA was administered. The normalized strain values (Fig. 8) before the change in flow rate did not deviate from those after. It seems that the alteration of flow rate had no effect. Six and 27 days, respectively, after the first addition of DCA, normal operation was slightly hampered temporarily. A permanent distortion occurred two days after the second addition. The total increase of strain was about the same as the increase found for Microcosm MES83-2.

Discussion

Evaluation of the test results was seriously hampered by the lack of information on the time course of localization and the actual concentration of DCA. However, transitions outside the normal operating range of the microcosms were found immediately after the addition of 10 μg L^{-1} DCA. In the most susceptible microcosm, this was the case within three days. DCA had to be administered in higher concentrations before it produced toxic effects on individual populations of algae, *Daphnia magna*, and *Hydra oligactis* in semichronic tests [12]. In these tests no toxic effects on *Daphnia magna* were found for concentrations up to 32 μg L^{-1}. The "no-observed-toxic-effect level" for the other organisms was even higher: 1000 μg L^{-1} DCA for the cyanobacterium *Microcystis aeruginosa* and 3200 μg L^{-1} for the green alga *Scenedesmus pannonicus* and for *Hydra oligactis*.

The use of dimethylsulfoxide (DMSO) as a solvent for DCA seemed possible on the basis of results of chronic toxicity tests in which DMSO proved to be totally harmless for three species of algae, three species of fish, and daphnids at the concentration of 0.1 mL L^{-1} [13]. As for DCA, effects of DMSO may occur at lower concentrations in the microcosms than in single-species

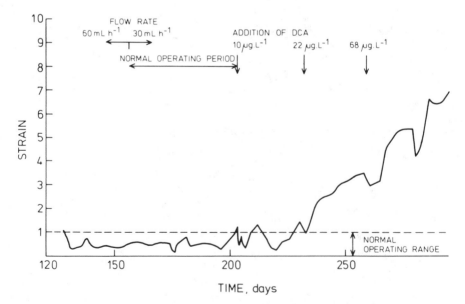

FIG. 8—*Time series of three-dimensional normalized strain in Microcosm MES85-2. The normal operating range is based on the biovolume of the algae, the number of daphnids, and the absolute difference of the pH values presented in Fig. 7 between Days 156 and 203. The moments of changing flow rate and toxicant addition are indicated.*

tests and may have influenced the observed changes. Synergistic toxic effects of DCA and DMSO cannot be excluded either.

The results suggest a difference between the reaction of Microcosms MES83-1 and MES83-3 and the reaction of the other two microcosms. The earlier transition of the state of Microcosms MES83-1 and MES83-3 outside the normal operating range can be interpreted as a smaller inertia of these microcosms. The smaller inertia cannot be the result of the manipulation of the flow rate and the predation on daphnids, because no manipulation (according to Table 1) appeared necessary before these microcosms exhibited values permanently outside their NORs. The greater increase of strain and the absence of temporary transitions in these two microcosms can be interpreted as a lower elasticity. This lower elasticity may be explained by the reduction in flow rate when a decrease in the number of daphnids was found as the result of the declining standing stock of the algae. The extra reduction in the outflow of algae to the daphnids may have accelerated the decrease in the number of dapnids. Since, after the addition of DCA, the polyps consumed less than the supplied amount of dapnids, the manipulation of the percentage of the population of daphnids that was transferred to the carnivore compartment cannot have influenced the response to DCA addition.

The differences in inertia and elasticity among the microcosms can also be explained by the stability of the microcosms during the normal operating range. The state variables of Microcosms MES83-2 and MES85-2 show greater variations during the normal operating period than the state variables of Microcosms MES83-1 and MES83-3. Owing to the greater operating range for these microcosms, the effect of DCA will have to be greater before a transition outside the normal operating range can be found. As a consequence, the observed transitions can only be interpreted in relation to the properties of the normal operating range.

A greater correlation between the state variables, and thereby a smaller normal operating range, can be expected when the phase differences between their fluctuations are taken into account [10]. For instance, a maximum in the density of algae probably caused a subsequent maximum in the number of herbivores. We have analyzed the transitions outside a normal operating range, using time shifts between the 1983 data series on algae, daphnids, and polyps, that gave the maximum correlation between these variables. The same differences in inertia and elasticity between the microcosms were found with this analysis as with the analysis just presented, without the time shifts. In this case, however, the change in the strain could not be interpreted without considering the changes in each state variable. A change in the strain could be the result of a change in the algal biovolume at Time t, or in the number of daphnids at Time t' (which equals t plus the time shift), or of the number of polyps at Time t''. Real-time simultaneous effects on more variables result in a partial change in strain until a time equal to the time shifts has passed and the changes due to all variables are reflected. Because it was felt that the supplementary information about the changes in the individual variables was not consistent with the aim of the statistical method used, this analysis with time shifts is not presented.

In real ecosystems, a decrease in the population size of the herbivores probably results in an increase in the standing stock of the algae grazed. In the microcosms, the only feasible way in which the grazing pressure on the autotrophic subsystem could be varied, along with the variation of the population size of the herbivores, seemed to be manipulation of the flow rate. However, this manipulation changed not only the outflow of algae from the autotrophic compartment, but also the inflow of nutrients into this compartment. As can be concluded from the absence of a change in the biovolume of algae after the reduction of the flow rate in the MES85-2 system, the algal standing stock was not affected. Contrary to a slowing down of the decrease in the number of daphnids, the reduced flow rate resulted in an extra reduction in the herbivores. For an increase in the algal standing stock in response to a decreasing herbivore population, the design of the microcosm needs further change. The outflow of the compartment with daphnids can be connected with the autotrophic compartment instead of the carnivore compartment. The chain of the carnivore and decomposer compartments should then be furnished with a second outflow of the algal compartment. In this way a loop for manipulation of the grazing process and a loop for manipulation of mineralization is created. The flexible modular design allows the construction of different systems according to the relative weight of isolated processes. Subsystems can be inserted or left out, and connections between subsystems and rates of processes can easily be changed. As has been proposed in the rationale for the use microcosms (Fig. 1), the physical design of a complex test system should not be standardized but needs careful and well-researched adaptation to the specific purposes of individual investigations.

Although the microcosms do not mimic natural ecosystems, they have proved to be valuable tools in assessing the effects of chemical perturbations on a community level of ecological organization. Hypotheses concerning these effects can be tested with these systems. In view of the "extrapolation problem," further research is needed on whether there is any regularity in microcosm response to different model compounds and whether effects observed in the type of microcosm used are representative of the effects to be expected in natural ecosystems. The microcosm proposed has the attractive property of developing a steady state. The steady-state properties make possible the test evaluation presented, which condenses the changes of several interdependent state variables into one, easily interpretable value proportional to the system strain. It appears to be a suitable method for the detection of changes in a relatively complex test system that cannot easily be replicated.

References

[1] Odum, H. T. and Hoskin, C. M., "Metabolism of a Laboratory Stream Microcosm," *Publications of the Institute of Marine Science, University of Texas,* Vol. 4, 1957, pp. 115-133.

[2] Giesy, J. P., Jr., and Odum, E. P. in *Microcosms in Ecological Research,* DOE Symposium Series 52, J. P. Giesy, Jr., Ed., National Technical Information Service, Springfield, VA, 1980, pp. 1-13.

[3] Cairns, J., Jr., in *Thermal Ecology II, Energy Research and Development Administration Symposium Series,* G. W. Esch and R. W. McFarlane, Eds., National Technical Information Service, Springfield, VA, 1976, pp. 32-38.

[4] Ringelberg, J. and Kersting, K., *Archiv für Hydrobiologie,* Vol. 83, 1978, pp. 47-68.

[5] Kersting, K., *Hydrobiological Bulletin,* Vol. 9, 1978, pp. 3-21.

[6] Kooijam, S. A. L. M., "De Dynamica van Populaties onder Chemische Stress," TNO-MT Rapport R 83/24, Netherlands Organization for Applied Scientific Research, Delft, The Netherlands, 1983.

[7] Abbott, W., *Journal of the Water Pollution Control Federation,* Vol. 38, 1966, pp. 258-270.

[8] Isensee, A. R., *International Journal for Environmental Studies,* Vol. 10, 1976, pp. 35-41.

[9] Vervelde, G. J. and Ringelberg, J., *Agro-Ecosystems,* Vol. 3, 1977, pp. 261-267.

[10] Kersting, K., *Internationale Revue der gesamten Hydrobiologie,* Vol. 69, 1984, pp. 567-607.

[11] Morrison, D. F., *Multivariate Statistical Methods,* McGraw-Hill, New York, 1976.

[12] Slooff, W. and Canton, J. H., *Aquatic Toxicology,* Vol. 4, 1983, pp. 271-282.

[13] Adema, D. M. M., Canton, J. H., Slooff, W., and Hanstveit, A. O., "Onderzoek Naar een Geschikte Combinatie Toetsmethoden ter Bepaling van de Aquatische Toxiciteit van Milieugevaarlijke Stoffen," TNO-MT Rapport CL 81/100, Netherlands Organization for Applied Scientific Research, Delft, The Netherlands, 1981.

Guy R. Lanza[1], *G. Allen Burton, Jr.,*[2] *and Joel M. Dougherty*[3]

Microbial Enzyme Activities: Potential Use for Monitoring Decomposition Processes

REFERENCE: Lanza, G. R., Burton, G. A., Jr., and Dougherty, J. M., **"Microbial Enzyme Activities: Potential Use for Monitoring Decomposition Processes,"** *Functional Testing of Aquatic Biota for Estimatin, Hazards of Chemicals, ASTM STP 988*, J. Cairns, Jr., and J. R. Pratt, Eds., American Society for Testing and Materials, Philadelphia, 1988, pp. 41–54.

ABSTRACT: Decomposition processes in subsurface soils and aquatic sediments have proven difficult to monitor because of their complex interactions. This research describes the use of indigenous microbial enzyme activities (MEA) and surrogate parameters to monitor decomposition in oligotrophic subsurface soils (76 to 168 cm below the surface) and in aquatic sediments from a eutrophic reservoir.

Measurements of total plate counts (TPC) of aerobic and anaerobic bacteria, alkaline phosphatase activity (APA), and total gas production in anaerobic subsurface soil microcosms are reported. Microcosms with titanium^{3+} citrate (TC) as a redox buffer appeared to have statistically higher numbers of anaerobes and to have uniform APA and gas production with and without sucrose amendment (nested ANOVA, $P < 0.001$). These studies indicate that subsurface soil microcosms have considerable potential for monitoring decomposition in vadose-zone soils.

Aquatic sediment microcosm studies revealed dose-response patterns in APA and dehydrogenase activity (DHA) after 96 h of exposure to As^{5+}, Cd^{2+}, Se^{4+}, and Cu^{2+}. The relative APA toxicities based on the calculated median effective concentration (EC_{50}) levels were $As^{5+} > Cu^{2+} > Cd^{2+} > Se^{4+}$. The relative DHA toxicities were the reverse: $Se^{4+} > Cd^{2+} > Cu^{2+} > As^{5+}$. As^{5+} produced a nonlinear dose-response pattern in DHA, which prevented calculation of the EC_{50} value. As^{5+} also produced decreased DHA activity at 10 mg/L and apparent stimulation of DHA at high levels, that is, 500 and 1000 mg/L. The proteolysis activity was highly variable (coefficient of variation greater than 40%) with five-day activities (percentage of controls) of 71 and 36% at 50 and 500-mg/L doses of As^{5+}, and 96 and 9% at the same doses of Cu^{2+}. No clear relationship was noted between MEA and either microbial density, as TPC, or acridine orange direct counts (AODC) with reduction of 2-iodophenyl-3-phenyl-5-nitrophenyl tetrazolium chloride (INT).

MEA appear useful in monitoring decomposition by indigenous microflora with and without toxicants. The MEA method detects sublethal indications of toxicant stress and can be correlated with other parameters commonly used to monitor ecosystem function.

KEY WORDS: hazard evaluation, microbial enzymes, microbial enzyme activity, subsurface soil bacteria, sediment bacteria, water pollution, median effective concentration (EC_{50}), metal toxicity, microbial microcosms, arsenic, selenium, copper, cadmium, decomposition processes

Microbial consortia in sediment and soil systems fill important niches in major decomposition processes. Natural and cultural activities influencing decomposition cycles are often reflected in sediment and soil microbial metabolism. Heterotrophic and biogeochemical production by microflora and biogeochemical cycling are essential for the utilization of substrates that

---

[1]Professor and chairman, Department of Environmental Health, East Tennessee State University, Johnson City, TN 37614.

[2]Assistant professor, Department of Biological Sciences, Wright State University, Dayton, OH 45435.

[3]Research assistant, Program in Environmental Sciences, University of Texas at Dallas, Richardson, TX 75083.

cannot be processed by animals and for the conversion of dissolved organic substrates to the particulate forms that are necessary links in natural food chains.

Interaction pathways in microbial consortia play important roles in the major functional processes of decomposition [1]. Interaction products (for example, total gas production or production of specific gases such as methane) can be used to monitor decomposition activity in either the absence or presence of pollutants. Broader measurements of microbial decomposition are obtainable by using enzyme activities common to microbial decomposers in sediment, soil, and water.

Microbial enzyme activities (MEA) can identify the relative decomposition of major substrates; for example, carbohydrates, proteins, and lipids in mixed-media organic matter cycling through ecosystems. Typical water values for carbohydrates, proteins, and lipids are 83.7, 15.6, and 0.7%, respectively. Within sediments, the organic matter decomposition rates follow this general sequence: carbohydrates, amino acids, amino sugars > humic compounds > lipids [2,3].

Microbial activities are commonly used to estimate overall community function, particularly the biomass [adenosine triphosphate (ATP) and lipids], growth rate, oxygen uptake [biological oxygen demand (BOD)], and carbon dioxide (CO_2) released [4-9]. Microbial enzyme activities have considerable potential for use in monitoring specific changes in heterotrophic decomposition processes. Table 1 [10-70] outlines selected microbial enzyme reactions available for such applications. Enzymes produced by microbial consortia catalyze many major transformation

TABLE 1—*Selected microbial enzyme reactions with potential for measuring decomposition activity and biomass in complex ecosystems.*

Activity Type (Enzyme Class)[a]	Major Habitat Tested	Selected References
Oxidoreductases (INT, TTC)	soil	[10-18]
	water	[19-24]
	sediment	[25-27]
Oxidases (glucose, amino acid polyphenol, ascorbate, urate)	soil	[28-35]
Catalases	soil	[17], [36-38]
	water	[39-40]
Peroxidases	soil	[41-43]
Transferases	soil	[44]
Hydrolases (lipases, esterases, phosphatases, ureases, proteinases)	soil (sludge)	[17], [45-49]
	water	[50-53]
	sediment	[54-58]
Enzyme complexes (total enzyme activity)		
Adenosine triphosphate (ATP-luciferase, GTP/ATP)	soil, water	[17], [59-63]
[¹³H]thymidine incorporation	sediment	[64]
[¹⁴C] and [³H] labeled solutes	sediment, water	[65-67]
Heat production	sediment	[68-70]
Gas production	subsurface soil	this report

[a]Key to abbreviations:
   INT = 2-Iodophenyl-3-phenyl-5-nitrophenyl tetrazolium chloride.
   TTC = 2,3,5-triphenyl-2H-tetrazolium chloride.
   GTP = guanosine triphosphate.

processes, and in some cases, microbes serve as the sole source of transformation enzymes in major food web processes [71]. Most of the enzyme activity types summarized in Table 1 can also be used to monitor decomposition processes in the presence of perturbations or pollutants or both. Microbial decomposition enzymes produced by consortia often respond in similar fashion to a wide array of environmental changes or pollutants, including uncouplers, inhibitors, toxins, and pH shifts [72].

Two of the authors of this paper recently reported the development of a battery of microbial enzyme activity tests suitable for use in complex environmental samples [73, 74]. The results, with and without model toxicants (As^{3+} and As^{5+}), were described for tests in water and aquatic sediments from lakes of varying trophic status. In this paper, we describe results of two additional approaches: (1) the development of a test system to measure selected decomposition activities of the indigenous microflora in microaerobic to anaerobic subsurface soils, and (2) the use of selected microbial enzyme activities (MEA) to monitor decomposition activities in aquatic sediments. The subsurface soil microcosms were used to monitor alkaline phosphatase activity (APA), total gas production, and microbial density under anaerobic conditions without toxicants. The aquatic sediment studies were done in microaerobic to anaerobic laboratory microcosms to monitor the effects of model toxicants on APA, dehydrogenase activity (DHA), proteolysis activity, and microbial density.

Materials and Methods

Subsurface Soils (Anaerobic)

Subsurface soils were collected from the Lyndon B. Johnson/Caddo National Grassland maintained by the U.S. Forest Service in Wise County, Texas.

Soil horizons were harvested with a mobile hydraulic drilling rig operated by the U.S. Soil Conservation Service. The following procedure was developed to collect near-subsurface cores (that is, 76 to 168 cm below the surface) with minimum exposure to aerobic conditions and surface microbial contamination. A hole was drilled with an alcohol-sterilized (70% ethanol) dry-tube core sampler approximately 5 cm in diameter and 60 cm in length. The core was quickly withdrawn and discarded. A second entry was made using an alcohol-sterilized dry-tube corer approximately 2.5 cm in diameter and 92 cm in length. These cores were immediately extruded into autoclaved anaerobe jars (Baltimore Biological Laboratories Gas Pak) receiving a continuous purge with zero-grade nitrogen. Approximately 2.5 cm of each end of each core was aseptically trimmed off just prior to its entry into an anaerobe jar. Gas Pak envelopes were activated to provide a maintenance atmosphere of hydrogen and carbon dioxide, and the jars were immediately sealed for transport to the laboratory. The anaerobiosis was monitored visually using in-jar methylene blue indicator strips. Soil Conservation Service personnel provided an on-site general description of the total soil horizons (from 0 to 168 cm below the surface) differentiating the major vertical zones and soil types.

Media and Reagents

All media and reagents were made using double-distilled and deionized water (Millipore). Casein-peptone-starch medium (CPS) [75] was used for heterotrophic plate counts. The CPS for anaerobic cultures was amended, after being autoclaved, with titanium^{3+} citrate (TC) at 192 mg/L Ti^{3+} as a reducing buffer [76]. Anaerobic CPS was made 24 h prior to use and stored anaerobically in Gas Pak jars (BBL) under a hydrogen and CO_2 atmosphere. The redox potential (Eh) of the anaerobically reduced CPS was -265 mV. The diluent for the plate counts consisted of a phosphate buffer [8.1 g of dibasic sodium phosphate (Na_2HPO_4) and 1.1 g of monobasic potassium phosphate (KH_2PO_4) per litre of water at pH 7.2] dispensed in screw-capped culture tubes (9.0 and 9.9 mL in final volume), which was autoclaved and then reduced with TC (91 and 100 mg/L as Ti^{3+}). The tubes were gassed with ultrahigh-purity (UHP) nitrogen using a Virginia Polytechnic Institute (VPI) anaerobic inoculator (Bellco), sealed, and stored at room temperature until needed.

The reconstituted soft water used to prepare the soil slurries consisted of Millipore H_2O containing 30 mg of calcium sulfate ($CaSO_4$), 30 mg of magnesium sulfate ($MgSO_4$), and 2 mg of potassium chloride (KCl) per litre [8]. The water was autoclaved, boiled, aseptically transferred to Wheaton 500-mL screw-capped bottles, and sealed under UHP nitrogen just prior to use.

Redox Buffer

The titanium citrate (TC) was made by modifying the original method of Zehnder and Wuhrmann [76]. Sterile Millipore H_2O in 350-mL volumes was amended with 100 mL of 1.03 M filter-sterilized trisodium citrate. This was brought to a boil, then cooled on ice while being gassed with UHP nitrogen. At room temperature, 38.75 mL of 20% titanium trichloride ($TiCl_3$) (Fisher stabilized reagent) was added. The pH was adjusted to 7.0 with 1.5 g of sodium carbonate (Na_2CO_3) and approximately 9 mL of 10 N sodium hydroxide (NaOH). The volume was adjusted to 500 mL with sterile Millipore water, and the solution was stored in screw-capped bottles sealed under UHP nitrogen. The Eh of these solutions was approximately -485 mV.

Soil Slurries

All manipulations involving the initial setup of the anaerobic soil slurry microcosms were done in a glove bag (Aldrich Co.) under UHP nitrogen. Subsurface soil cores were retrieved from storage in Gas Pak jars and homogenized in a Waring blender with and without TC. The cores were blended until the homogenized soil was of uniform color. The homogenates were stored in the Gas Pak jars until used in experiments.

Soil slurries were made by blending modified reconstituted water with homogenized soil 1:1 in a Waring blender until a stable creamy consistency was obtained. A 60-mL plastic syringe was used to transfer 100 mL of this slurry (in 50-mL portions) to each microcosm. The slurries were contained in 250-mL glass bottles (Wheaton) with plastic caps fitted with butyl rubber septums. These bottles were gassed out before the addition of the slurries. After addition of 100 mL of slurry to each bottle, 150 mL of modified reconstituted soft water was added, and the units were sealed.

Microcosm Groups

Four groups, each with triplicate microcosms, were established and incubated anaerobically on gyroshakers at 200 rpm for nine months. The groups were operated as follows: Group 1, with sucrose and without TC; Group 2, without sucrose or TC; Group 3, with sucrose and TC; Group 4 without sucrose and with TC. Sucrose was added three weeks after the microcosm start-up to yield a final concentration of 1600 mg sucrose/L.

Physical and Chemical Measurements

The slurry aliquots were removed aseptically using a sterile syringe, placed in 15-mL conical centrifuge tubes, and gassed with UHP nitrogen. Eh and pH measurements were taken with a combination meter (Corning Model 140). The Eh was measured by immersing a micro-redox probe in the slurry and allowing a 4-min equilibration period before recording the results. The pH was taken with a combination probe. Gas production was measured volumetrically by displacement in a 10-mL syringe and tabulated over the total incubation period. The Eh, pH, and gas production were monitored at 24 to 72-h intervals during the first month of microcosm operation and were checked one month before the end of the operation period.

Microbial Growth and APA Activity

Two weeks before the end of the nine-month anaerobic incubation, slurry aliquots were taken aseptically with a sterile syringe for measurements of microbial density and APA. Total plate counts (TPC) were prepared as spread plates on appropriate media for aerobic and anaerobic bacteria. The anaerobic dilutions were done with a VPI anaerobic inoculator (Bellco Glass Co.), and the plates were kept in Gas Pak jars under hydrogen and carbon dioxide. All plates were incubated for ten days at 22°C and then counted.

APA was determined spectrophotometrically [73] by measuring p-nitrophenol (p-NP) cleavage from p-nitrophenyl phosphate (p-NPP) (Sigma Chemical Co.).

Statistical Precision

Statistical analyses were performed on the IBM computer at the University of Texas at Dallas using a canned statistical program [77]. Untransformed data were entered into the SAS general linear models procedure (PROC GLM) for both single-classification analysis of variance (ANOVA) and nested designs. Raw data were examined for normality of distribution and homogeneity of variance, and appropriate transformations were made as needed [78]. Significant differences revealed by ANOVA were isolated by the Waller-Duncan multiple comparison of means. In nested designs, PROC NESTED was used to determine the percentage of variance contributed by each level in the analysis. TPC were done using three-level nested ANOVA, gas production and APA analyses using single-classification ANOVA.

Aquatic Sediments

Aquatic sediments were collected from Lake Lavon, a eutrophic reservoir in Collin County, Texas, and were used to establish laboratory microcosms as described previously [73].

Microcosm Groups

Replicate microcosms for APA and DHA were dosed with As^{5+}, Cd^{2+}, Se^{4+}, and Cu^{2+} at 10, 100, 500, and 1000 mg/L and were incubated at 25°C for 96 h. The microcosms for the proteolysis study were dosed with As^{5+} and Cu^{2+} at two levels, 50 and 500 mg/L, and were incubated for five days.

Microbial Enzyme Activities and Microbial Density

APA (p-NP cleavage), DHA [triphenyltetrazolium chloride (TTC)], proteinase [hide powder azure (HPA) degradation], total plate counts (TPC), and acridine orange direct counts (AODC) with reduction of 2-iodophenyl-3-phenyl-5-nitrophenyl tetrazolium chloride (INT) were examined using methods described previously [73]. The median effective concentration (EC_{50}) values were computed using SAS probit analysis [78].

Results and Discussion

Soil Slurries

The microcosm Eh ranges, in millivolts, fluctuated over the study period as follows: Group 1, 145 to −246 (24 h after sucrose addition); Group 2, 172 to −46; Group 3, 185 to −125; and Group 4, 189 to −206 mV. The microcosm pH ranges over the same period showed a slight trend toward decreasing values: for Group 1, the pH was 7.04 to 6.53; for Group 2, 7.15 to 6.47; for Group 3, 6.70 to 6.24; and for Group 4, 6.70 to 6.53.

Figure 1 and Table 2 provide a statistical summary of the microcosm microbial performance over the nine-month anaerobic study period.

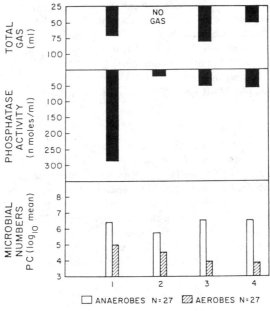

MICROCOSM GROUP
Group 1 : No Ti (Ⅲ), Sucrose-amended
Group 2 : No Ti (Ⅲ), No Sucrose
Group 3 : Ti (Ⅲ), Sucrose-amended
Group 4 : Ti (Ⅲ), No Sucrose

FIG. 1—*Statistical summary of the subsurface soil microcosm performance over the nine-month anaerobic study period (nested ANOVA, P < 0.001).*

TABLE 2—*Statistical summary of the subsurface soil microcosm performance over the nine-month anaerobic study period (see Fig. 1) using nested ANOVA.*[a]

Parameter	Microcosm Group
Anaerobic bacteria ($\log_{10}$)	1 2 <u>3</u> <u>4</u> (highest)
Aerobic bacteria ($\log_{10}$)	1 2 <u>3</u> <u>4</u> (lowest)
Phosphatase enzyme activity	1 2 <u>3</u> <u>4</u>[b]
Total gas production	<u>1</u> 2 <u>3</u> 4

[a]The underlined groups are not significantly different.
[b]$P < 0.05$.

The anaerobic microbial activity spectrum was statistically higher in Groups 3 and 4, which were prepared with the TC redox buffer, although Group 1 microcosms (without TC) had the highest levels of APA decomposition activity and the same levels of gas produced as Group 3. However, the highest levels of anaerobic bacteria and the lowest of aerobic bacteria occurred in the groups with TC, Groups 3 and 4. No detectable gas production occurred in Group 2 (without TC) unless it was amended with sucrose. The phosphatase activity was uniform in Groups 3 and 4 (with no statistically significant differences). Group 3 had higher gas production than Group 4 (without sucrose). Qualitative observations of anaerobic TPC from TC-treated anaerobic microcosms indicated greater organism diversity, based on colony morphology and pigmentation.

These studies indicate that subsurface soil microcosms have considerable potential as functional tools for monitoring vadose-zone decomposition activities. The statistical framework permits comparison of microbial density, phosphatase activity, and total gas production under different operating regimes. Microcosms constructed to reflect selected decomposition cycles involving indigenous microbial consortia under anaerobic conditions should be explored further. Microcosms similar to those described here could provide additional information on the effects of soil pollutants percolating to vadose compartments.

Aquatic Sediments

The results of 96-h Lake Lavon sediment microcosm studies are summarized in Figs. 2 through 5 and in Table 3 [*79–82*]. Arsenate (As^{5+}) produced a dose-response pattern in APA with no additional decrease at concentrations above 500 mg/L (Fig. 2) and with a calculated [*77*] EC_{50} of 64 mg/L (Table 3). DHA showed a variable and split response with a precipitous

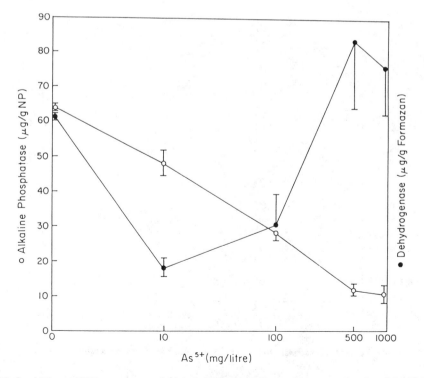

FIG. 2—*APA and DHA responses to As^{5+} in Lake Lavon sediment microcosms incubated for 96 h.*

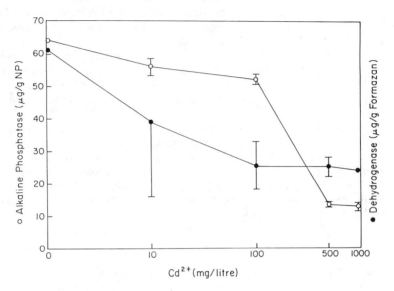

FIG. 3—*APA and DHA responses to Cd^{2+} in Lake Lavon sediment microcosms incubated for 96 h.*

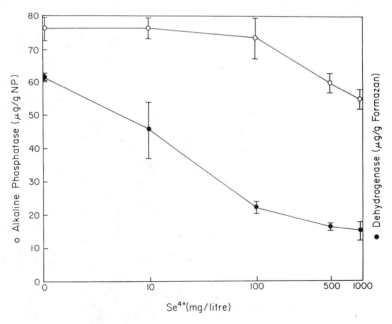

FIG. 4—*APA and DHA responses to Se^{4+} in Lake Lavon sediment microcosms incubated for 96 h.*

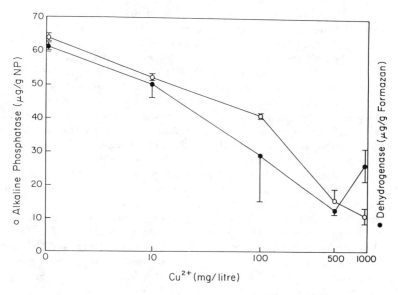

FIG. 5—*APA and DHA responses to Cu^{2+} in Lake Lavon sediment microcosms incubated for 96 h.*

TABLE 3—*Inhibition of microbial enzyme activities in selected microbial populations.*

	EC_{50}, mg/L[a]			
	This Report		Other Reports[b]	
Toxicant	DHA	APA	DHA	APA
As^{5+}	[c]	64	...	...
Cd^{2+}	76	197	3.3 [79][d]	...
			1.8 [80][e]	
			20 [82][f]	
Se^{4+}	60	...	...	...
Cu^{2+}	136	123	3.0 [79][d]	
			12.3 [80][e]	...
			0.59 [81][f]	
			53 [82][g]	

[a]The 96-h EC_{50} in this report was calculated using SAS probit analysis [77].
[b]The brackets contain references.
[c]Not calculated due to nonlinear dose-response results.
[d]Pure cultures of *Escherichia coli.*
[e]Pure cultures of *Pseudomonas alcaligenes,* INT-dehydrogenases.
[f]Mean inhibition concentration (IC_{50}) of raw sewage communities, INT-dehydrogenases.
[g]Soil microcosms.

decline at 10 mg/L As^{5+} and a lesser decline in activity at 100 mg/L. The observed nonlinear dose-response results for arsenate did not permit the calculation of an EC_{50}.

In an earlier report [73], two of this paper's authors noted a similar nonlinear DHA response to As^{3+} (arsenite) by sediment consortia. Other investigators have described incomplete inhibition of DHA in susceptible soil consortia, accompanied by no inhibition of a resistant consortium in the same soil [82]. Differential DHA at different metal concentrations could reflect either a mixed-population response (that is, consortia with both susceptible and resistant microbes) or a change in the metabolic response as a result of changing levels of a metal (for example, arsenate, acting as an uncoupler, can produce stimulated electron transfer activity). Finally, it is interesting to note the similarity of the DHA response to As^{5+} in our sediment microcosms to that reported in a Microtox assay (94 mg/L As^{5+}) [83].

Figure 3 summarizes the APA and DHA responses to Cd^{2+}. The APA declined slightly to levels of 100 mg/L Cd^{2+}, with a sharp decrease at higher toxicant levels. The calculated EC_{50} (Table 3) was 197 mg/L. The DHA followed a dose-response pattern to 100 mg/L Cd^{2+}, with a uniform response at higher toxicant levels. The DHA EC_{50} for Cd^{2+} was 76 mg/L.

Microbial response patterns for Cd^{2+} have been extensively studied [84]. Cd^{2+} has been reported to elicit varying growth rate responses in soil consortia (bacteria and actinomycetes), with the organisms exhibiting tolerant, intolerant, and stimulated behavior. One community was observed to be 50% intolerant to 5 mg/L Cd^{+2} but tolerant to levels above 5 mg/L [85]. Reported Cd^{2+} EC_{50} values for DHA (Table 3) ranged between 1.8 and 20 mg/L. Lower values (1.8 and 3.3 mg/L) were obtained from pure cultures, while the higher value (20 mg/L) was found in soil under conditions closer in complexity to those of aquatic sediments.

Microbial APA in response to selenite (Se^{4+}) indicated no detectable effect for levels reaching 100 mg/L, but some depression of activity at 500 to 1000 mg/L (Fig. 4). DHA followed a dose-response pattern for concentrations up to 500 mg/L, with no additional depression at 1000 mg/L. The EC_{50} for DHA was 60 mg/L. Of the metals reported here, Se^{4+} produced the highest DHA and lowest APA toxicity in aquatic sediment microcosms.

APA followed a general response to Cu^{2+} with an apparent uniform effect above 500 mg/L. DHA followed a similar pattern with evidence of recovery or stimulation or both at 1000 mg/L. The literature EC_{50} for DHA exposed to Cu^{2+} (Table 3) ranged from 0.59 to 53 mg/L. As was noted with Cd^{2+}, the lower levels (3.0 and 12.3 mg/L Cu^{2+}) resulted from tests done over short incubation periods with pure cultures of bacteria. One exception was a study done with microbial consortia from raw sewage, using INT/dehydrogenase activity to estimate a mean inhibition concentration (IC_{50}) of 0.59 mg/L Cu^{2+} [81]. The highest literature Cu^{2+} EC_{50} in Table 3 is 53 mg/L, generated in a soil study more closely approximating the complex variables in our sediment microcosms.

Proteolysis

Proteolysis activity was highly variable, with a coefficient of variation (CV) typically greater than 40%. The activity values (percentage of control activity) were 71 and 36% at 50 and 500 mg/L doses of As^{5+}, and 96 and 9% at 50 and 500 mg/L doses of Cu^{2+}. High variability in proteolysis activity has been reported in water [86] and in sediment [73].

Microbial Density

No clear patterns or relationships between MEA and either TPC or AODC (INT) were noted. The percentage of active bacteria (INT) was extremely variable in both the control and toxicant-dosed sediments. Lack of correlation between TPC, AODC, and microbial biomass has been reported in other microcosm studies [87]. Refinement of existing methods for measuring biomass and activity may improve techniques currently in use [88]. Past studies have often ignored major methodological variables affecting AODC and AODC (INT) measurements. Some of these variables and interferences were recently clarified [73,74] and should be considered in future studies of complex environmental systems.

Summary and Conclusions

Microbial enzyme activities (MEA) have considerable potential for describing selected aspects of soil/sediment/water decomposition processes. MEA can be used in the absence or presence of pollutants. We have examined APA, DHA, and proteolysis because of their ubiquitous role in decomposition events and their adaptability to use in complex environmental samples (subsurface soils, aquatic sediments, and water).

Typical organic matter is largely phosphorylated organic material, mixed carbohydrates and proteins (including amino acids and amino sugars); APA, DHA, and proteolysis are well-suited to monitor their decomposition. APA is usually induced when the available phosphate declines to low levels [89], DHA follows respiratory electron transport activity during many common decomposition activities [90], and proteolysis reflects the protein solubilization process through hydrolysis of peptide linkages to release peptones, peptides, and amino acids [48].

There are several distinct advantages in using MEA to monitor decomposition processes. First, the approach permits the description of different aspects of organic matter breakdown in a microcosm over similar time frames (for example, respiratory activity, phosphorus cleavage from organic molecules, and protein breakdown). Second, MEA offer a more specific measurement of microbial function than other more generic tests, such as BOD or ATP. MEA describe a broad spectrum of microbial metabolic types and reflect both the eucaryotic and procaryotic species components making up typical organism consortia. This is especially important in monitoring decomposition processes over seasonal and yearly cycles. For example, typical subsurface soil systems and their microbial consortia are largely oligotrophic, receiving pulses of organic matter derived from surface sources. Aquatic sediments are also pulsed with organic matter derived from autochthonous (for example, phytoplankton blooms) and allochthonous (for example, terrestrial detritus) organic matter. These organic matter fluxes drive and regulate a changing decomposition regime not easily characterized by nonspecific tests. MEA could clarify increases and decreases in each enzymatic pathway over seasonal cycles.

Finally, APA, DHA, and proteolysis appear to be good candidates as sublethal indicators of microbial decomposition activities challenged by the appearance of toxicants. Total suppression of DHA by chlorination even while considerable numbers of bacteria remained viable has been reported [91]. In these studies, toxicants suppressed each MEA component with different dose-response patterns but never caused complete enzymatic suppression. This suggests that even high levels of some toxicants may not cause the death of all decomposer organisms in a consortium.

Currently, MEA are most useful in monitoring undisturbed decomposer consortia and those stressed by toxicants. Both applications are important since the integrity of the ecosystem depends heavily on the performance of decomposer consortia. In these studies, subsurface soil microcosms with indigenous consortia made it possible to monitor APA, total gas production, and microbial density with good statistical precision. Sediment microcosms produced responses to As^{5+}, Cd^{2+}, Se^{4+}, and Cu^{2+}, with the relative toxicities to APA being $As^{5+} > Cu^{2+} > Cd^{2+} > Se^{4+}$ and to DHA being $Se^{4+} > Cd^{2+} > Cu^{2+}$, As^{5+}. The reverse toxicity order noted between the two MEA underscores the need for more than one functional indicator. The proteolysis activity was highly variable with CV > 40%. TPC and AODC (INT) did not correlate with MEA.

Future research on MEA as a functional test should clarify the following: (1) the interactions between individual enzyme activities, for example, the amount of phosphatase activity required to provide the biologically available phosphorus necessary to support proteolysis, and (2) the relationship between MEA and other ecological parameters commonly used to characterize soils, sediments, and water, for example, correlations between respiration and DHA and between carbon/nitrogen/phosphorus ratios and phosphatase activity. We recently reported statistically significant correlations between MEA and response studies of in-stream biological communities exposed to complex wastewater effluents [92], which further demonstrates the sensitivity and usefulness of MEA in environmental studies.

Acknowledgments

Part of this research was funded by the U.S. Environmental Protection Agency, Office of Research and Development (Grant No. R81134-01). We are grateful to Al Scott and Dennis Clower, U.S. Soil Conservation Service, for help with soil collection and characterization; Ben Harbor, U.S. Forest Service, for access to the soil study site; and Judy Welton for assistance with the manuscript.

References

[1] Lanza, G. R. and Silvey, J. K. G. in *Microbial Processes in Reservoirs*, D. Gunnison, Ed., W. Junk, Boston, 1985, Chapter 6, pp. 99–119.

[2] Kemp, A. L. W. and Johnson, L. M., *Journal of Great Lakes Research*, Vol. 5, 1979, pp. 1–10.

[3] Wetzel, R. G., *Limnology*, 2nd ed., Saunders, Philadelphia, 1983, pp. 592–593.

[4] Karl, D. M. and LaRock, P. A., *Journal of the Fisheries Research Board of Canada*, Vol. 32, 1975, pp. 599–607.

[5] Federle, T. W. and White, D. C., *Applied and Environmental Microbiology*, Vol. 44, 1982, pp. 1166–1169.

[6] White, D. C., Davis, W. M., Nickels, J. S., King, J. D., and Bobbie, R. J., *Oecologia*, Vol. 40, pp. 51–62.

[7] White, D. C., Bobbie, R. J., Nickels, J. S., Fazio, S. D., and Davis, W. M., *Botanica Marina*, Vol. 23, pp. 239–250.

[8] *Standard Methods for the Examination of Water and Wastewater*, 14th ed., American Public Health Association, Washington, DC, 1976, pp. 543–550.

[9] Sorokin, Y. I. and Kadota, H., *Techniques for the Assessment of Microbial Production and Decomposition in Fresh Waters*, IBP Handbook No. 23, Blackwell Scientific Publications, Oxford, Great Britain, 1972, pp. 15–39.

[10] Stevenson, I. L., *Canadian Journal of Microbiology*, Vol. 5, 1959, pp. 229–235.

[11] Kiss, S. and Boaru, M., "Methods for the Determination of Dehydrogenase Activity in Soil," *Proceedings*, Symposium on Methods for Soil Biology, Bucharest, Romania, 1965, pp. 137–143.

[12] Lenhard, G., *Soil Science*, Vol. 101, 1966, pp. 400–402.

[13] Bremmer, J. M. and Tabatabai, M. A., *Soil Biology and Biochemistry*, Vol. 5, 1973, pp. 385–386.

[14] Skujins, J. J., *Bulletin of the Ecological Research Commission* (Stockholm), Vol. 17, 1973, pp. 235–241.

[15] Domsch, K. H., Beck, T., Anderson, J. P. E., Sonderstrom, B., Parkinson, D., and Trolldenier, G., *Zeitschrift fuer Pflanzenernaehrung und Bodenkunde*, Vol. 142, 1979, pp. 520–533.

[16] Trevors, J. T., Mayfield, C. I., and Inniss, W. E., *Microbial Ecology*, Vol. 8, 1982, pp. 163–168.

[17] Casida, L. E., *Applied and Environmental Microbiology*, Vol. 34, 1977, pp. 630–636.

[18] Klein, D. A., Sorenson, D. L., and Redente, E. F. in *Soil Reclamation Processes, Microbiological Analyses and Applications*, R. L. Tate and D. A. Klein, Eds., Marcel Dekker, New York, 1985, Chapter 5, pp. 141–171.

[19] Curl, H., Jr., and Sandberg, J., *Journal of Marine Research*, Vol. 19, 1961, pp. 123–138.

[20] Packard, T. T. and Healy, M. L., *Journal of Marine Research*, Vol. 26, 1968, pp. 66–74.

[21] Packard, T. T., *Journal of Marine Research*, Vol. 29, 1971, pp. 235–244.

[22] Zimmermann, R., Iturriaga, R., and Becker-Birck, J., *Applied and Environmental Microbiology*, Vol. 36, 1978, pp. 926–935.

[23] Maki, J. S. and Remsen, C. C., *Applied and Environmental Microbiology*, Vol. 41, 1981, pp. 1132–1138.

[24] Fliermans, C. B., Bettinger, G. E., and Fynskk, A. W., *Water Research*, Vol. 16, 1982, pp. 903–909.

[25] Zimmerman, A. P., *Internationale Vereinigung für Theoretische und Angewandte Limnologie*, Vol. 19, 1975, pp. 1518–1523.

[26] Iturriaga, V. and Rheinheimer, G., *Kieler Meeresforschungen*, Vol. 31, 1975, pp. 83–86.

[27] Bright, J. J. and Fletcher, M., *Applied and Environmental Microbiology*, Vol. 45, 1983, pp. 818–825.

[28] Durand, G., *Annales de l'Institut Pasteur*, Vol. 107, Supplement, 3, 1964, pp. 136–147.

[29] Ross, D. J., *Plant Soil*, Vol. 28, 1968, pp. 1–11.

[30] Kuprevich, V. F. and Shcherbakova, T. A. in *Soil Biochemistry*, 2nd ed., A. D. McLaren and J. J. Skujins, Eds., Marcel Dekker, New York, 1971, pp. 167-201.
[31] Ross, D. J. and McNeilly, B. A., *Soil Biology and Biochemistry*, Vol. 4, 1972, pp. 9-18.
[32] Bordeleau, L. M. and Bartha, R., *Canadian Journal of Microbiology*, Vol. 18, 1972, pp. 1865-1871.
[33] Ross, D. J. and McNeilly, B. A., *New Zealand Journal of Science*, Vol. 16, 1973, pp. 241-257.
[34] Ross, D. J., *Soil Biology and Biochemistry*, Vol. 6, 1974, pp. 303-306.
[35] Lowery, S. N., Carr, P. W., and Seitz, W. R., *Analytical Letters*, Vol. 10, 1977, pp. 931-943.
[36] Weetall, H. H., Weliky, N., and Vango, S. P., *Nature*, Vol. 206, 1965, pp. 1019-1021.
[37] Skujins, J. J. and McLaren, A. D., *Enzymologia*, Vol. 34, 1968, pp. 213-225.
[38] Kunze, C., *Experientia*, Vol. 28, 1972, pp. 1397-1398.
[39] Sakaguchi, B., *Japan Analyst*, Vol. 2, 1953, p. 226.
[40] Stefanic, G. and Dumitru, L., *Biological Science*, Vol. 12, 1970, pp. 12-13.
[41] Bartha, R. and Bordeleau, L. M., *Soils and Fertilizers*, Vol. 33, 1969, p. 1448.
[42] Lay, M. M. and Illnicki, R. D., *Weed Research*, Vol. 14, 1974, pp. 111-113.
[43] Puget, K., Michelson, A. M., and Avrameas, S., *Analytical Biochemistry*, Vol. 79, 1977, pp. 447-456.
[44] Skujins, J. J. in *Soil Biochemistry*, 1st ed., A. D., McLaren and G. H. Peterson, Eds., Marcel Dekker, New York, 1967, pp. 371-414.
[45] Ingols, R. S., "A Study of Hydrolytic Enzymes in Activated Sludge," *Bulletin of the New Jersey Agricultural Experiment Station*, No. 669, New Brunswick, NJ, 1939.
[46] Halstead, R. L., *Canadian Journal of Soil Science*, Vol. 44, 1964, pp. 137-144.
[47] Ramirez-Martinez, J. R. and McLaren, A. D., *Enzymologia*, Vol. 31, 1966, pp. 23-38.
[48] Tabatabai, M. A. and Bremmer, J. M., *Soil Biology and Biochemistry*, Vol. 1, 1969, pp. 301-307.
[49] Spier, T. W. and Ross, D. J., *New Zealand Journal of Science*, Vol. 18, 1975, pp. 231-237.
[50] Gilmartin, M., *Limnology and Oceanography*, Vol. 12, 1967, pp. 325-328.
[51] Perry, M. J., *Marine Biology*, Vol. 15, 1972, pp. 113-119.
[52] Jansson, M., *Science*, Vol. 194, 1976, pp. 320-321.
[53] Little, J. E., Sjogren, R. E., and Carson, G. R., *Applied and Environmental Microbiology*, Vol. 37, 1979, pp. 900-908.
[54] Sizemore, R. K., Stevenson, L. H., and Hebeler, B. H. in *Estuarine Microbial Ecology*, 1st ed., L. H. Stevenson and R. R. Colwell, Eds., University of South Carolina Press, Columbia, SC, 1973, pp. 133-141.
[55] Flint, K. P. and Hopton, J. W., *European Journal of Applied Microbiology*, Vol. 4, 1977, pp. 204-215.
[56] White, D. C., Bobbie, R. J., Herron, J. S., King, J. D., and Morrison, S. J., "Biochemical Measurements of Microbial Biomass and Activity from Environmental Samples," *Native Aquatic Bacteria: Enumeration, Activity, and Ecology, ASTM STP 695*, American Society for Testing and Materials, Philadelphia, 1979.
[57] Stewart, A. J. and Wetzel, R. G., *Freshwater Biology*, Vol. 12, 1982, pp. 369-380.
[58] Caplan, J. A. and Fahey, J. W., *Bulletin of Environmental Contamination and Toxicology*, Vol. 25, 1980, pp. 424-426.
[59] Geesey, G. G. and Costerton, J. W., "Bacterial Biomass Determinations in a Silt-Laden River: Comparison of Direct-Count Epifluorescence Microscopy and Extractable Adenosine Triphosphate Techniques," *Native Aquatic Bacteria: Enumeration, Activity, and Ecology, ASTM STP 695*, American Society for Testing and Materials, Philadelphia, 1979.
[60] Stevenson, L. H., Chrzanowski, T., and Erkenbrecher, C. W., "The Adenosine Triphosphate Assay: Conceptions and Misconceptions," *Native Aquatic Bacteria: Enumeration, Activity, and Ecology, ASTM STP 695*, American Society for Testing and Materials, Philadelphia, 1979.
[61] Karl, D. M., "Adenosine Triphosphate and Guanosine Triphosphate Determinations in Intertidal Sediments," *Methodology for Biomass Determinations and Microbial Activities in Sediments, ASTM STP 673*, American Society for Testing and Materials, Philadelphia, 1979.
[62] Karl, D. M., *Microbiological Reviews*, Vol. 44, 1980, pp. 739-796.
[63] Deming, J. W., Picciolo, G. L., and Chappelle, E. W., "Important Factors in Adenosine Triphosphate Determinations Using Firefly Luciferase: Applicability of the Assay to Studies of Native Aquatic Bacteria," *Native Aquatic Bacteria: Enumeration, Activity, and Ecology, ASTM STP 695*, American Society for Testing and Materials, Philadelphia, 1979.
[64] Newell, S. Y. and Fallon, R. D., *Microbial Ecology*, Vol. 8, 1982, pp. 33-46.
[65] Wright, R. T., *Applied and Environmental Microbiology*, Vol. 36, 1978, pp. 297-305.
[66] Meyer-Reil, L. A., *Applied and Environmental Microbiology*, Vol. 36, 1978, pp. 506-512.
[67] Ladd, T. I., Costerton, J. W., and Geesey, G. G., "Determination of Heterotrophic Activity of Epilithic Microbial Populations," *Native Aquatic Bacteria: Enumeration, Activity, and Ecology, ASTM STP 695*, American Society for Testing and Materials, Philadelphia, 1979.
[68] Pamatmat, M. M. and Bhagwat, A. M., *Limnology and Oceanography*, Vol. 18, 1973, pp. 611-627.
[69] Pamatmat, M. M., Bengtsson, W., and Novak, C. S., "Heat Production, ATP Concentration and

Electron Transport Activity of Marine Sediments," Marine Ecology Program Series 4, Department of Oceanography, University of Washington, Seattle, WA, 1981.

[70] Pamatmat, M. M., *Science*, Vol. 215, 1982, pp. 395-396.

[71] Griffiths, R. P., *Marine Pollution Bulletin*, Vol. 14, 1983, pp. 162-165.

[72] Broda, E., *The Evolution of the Bioenergetic Processes*, Pergamon Press, New York, 1978, p. 96.

[73] Burton, G. A., Jr., and Lanza, G. R., "Sediment Microbial Activity Tests for the Detection of Toxicant Impacts," *Aquatic Toxicology and Hazard Assessment: Seventh Symposium, ASTM STP 854*, American Society for Testing and Materials, Philadelphia, 1985.

[74] Burton, G. A., Jr., and Lanza, G. R., *Applied and Environmental Microbiology*, Vol. 51, 1986, pp. 931-937.

[75] Collins, V. G. and Willoughby, L. G., *Archives für Microbiologie*, Vol. 43, 1962, pp. 294-307.

[76] Zehnder, A. J. and Wuhrmann, K., *Science*, Vol. 194, 1976, pp. 1165-1166.

[77] Ray, A. A. in *SAS Users Guide: Statistics*, SAS Institute, Inc., Cary, NC, 1982, Chapters 7, 9, 10, and 18, pp. 113-204 and Probit Analysis, Chapter 18, pp. 287-294.

[78] Rohlf, F. J. and Sokal, R. R., *Biometry*, W. H. Freeman, San Francisco, 1969, Chapters 8-11, pp. 175-353.

[79] Cenci, G., Morozzi, G., and Caldini, G., *Bulletin of Environmental Contamination and Toxicology*, Vol. 34, 1985, pp. 188-195.

[80] Bitton, G., Khafif, T., Chalaigner, N., Bastide, J., and Coste, C. M., *Toxicity Assessment: An International Quarterly*, Vol. 1, 1986, pp. 1-12.

[81] Dutton, R. J., Bitton, G., and Koopman, B., *Toxicity Assessment: An International Quarterly*, Vol. 1, 1986, pp. 147-158.

[82] Rogers, J. E. and Li, S. W., *Bulletin of Environmental Contamination and Toxicology*, Vol. 34, 1985, pp. 858-865.

[83] Chang, J. C., Taylor, P. B., and Leach, F. R., *Bulletin of Environmental Contamination and Toxicology*, Vol. 26, 1981, pp. 150-156.

[84] Trevors, J. T., *Canadian Journal of Microbiology*, Vol. 32, 1986, pp. 447-464.

[85] Williams, S. E. and Wollum, A. G. II, *Journal of Environmental Quality*, Vol. 10, 1981, pp. 142-144.

[86] Erstraete, V., Voets, J. P., and van Lancker, P., *Hydrobiologia*, Vol. 49, 1976, pp. 257-266.

[87] Domsch, K. H., Beck, T., Anderson, J. P. E., Sonderstrom, B., Parkinson, D., and Trolldenier, G., *Zeitschrift fuer Pflanzenernaehrung und Bodenkunde*, Vol. 142, 1979, pp. 520-533.

[88] Paul, E. A. and Voroney, R. P. in *Current Perspectives in Microbial Ecology*, M. J. Klug and C. A. Reddy, Eds., American Society for Microbiology, Washington, DC, 1984, pp. 509-514.

[89] Hashimoto, S., Fujiwara, K., and Fuwa, K., *Journal of Environmental Science and Health*, Vol. A20, 1985, pp. 781-809.

[90] Trevors, J. T., *Water Research*, Vol. 18, 1984, pp. 581-584.

[91] Roller, S. D., Olivieri, V. P., and Kawata, K., *Water Research*, Vol. 14, 1980, pp. 635-641.

[92] Burton, G. A., Jr., and Lanza, G. R., *Water Research*, Vol. 21, 1987, pp. 1173-1182.

James R. Pratt[1] *and John Cairns, Jr.*[2]

Use of Microbial Colonization Parameters as a Measure of Functional Response in Aquatic Ecosystems

REFERENCE: Pratt, J. R. and Cairns, J., Jr., "**Use of Microbial Colonization Parameters as a Measure of Functional Response in Aquatic Ecosystems,**" *Functional Testing of Aquatic Biota for Estimating Hazards of Chemicals, ASTM STP 988,* J. Cairns, Jr., and J. R. Pratt, Eds., American Society for Testing and Materials, Philadelphia, 1988, pp. 55-67.

ABSTRACT: Microbial colonization of artificial substrates introduced into aquatic ecosystems is affected by the relative levels of nutrients and toxicants. The productivity of microbial biota integrates several factors affecting organism survival and is expressed in relative rates of propagule production and colonization. It is now possible to examine factors affecting microbial colonization in laboratory microcosms. Colonized artificial substrate species sources can be used in the laboratory to measure nutrient and toxicant effects. Studies have shown that microbial colonization is at least as sensitive a technique as long-term single-species testing and has allowed controlled measurement of complex community responses to disturbance. It is possible to model the nonlinear colonization process and to compare colonization rates, equilibrium species numbers, and biomass production in test systems. Such testing can examine effects of nutrients, pure toxicants, or complex effluents. The authors found adverse effects on species dispersal in static test systems for cadmium and copper at concentrations of <1 μg Cd/L and 18 μg Cu/L, although low levels of copper enhanced the species numbers. In microcosm systems receiving continuous toxic input, we found adverse effects of chlorine on species dispersal at 2 μg/L; however, net production was elevated at concentrations up to 100 μg/L. In tests with the herbicide atrazine, the number of species and the net production were stimulated at low levels (3 to 30 μg/L). Stimulation of the species number and production may be a result of effects on control mechanisms and not the result of a subsidy to the community. The study of natural community dynamics in evaluating the effects of toxic materials provides evidence of effects on the emergent properties of systems, which are not available from studies of individual species. In many cases, effects on communities may occur at concentrations similar to those producing effects on sensitive single species. However, testing at the community or system level can provide predictions of community or system components most at risk. As such, system level tests should provide predictions that can be directly validated in ecosystems receiving toxic inputs.

KEY WORDS: hazard evaluation, microbial colonization, microbial communities, artificial substrates, microcosms, toxicity, microbial production, protozoa, heavy metals, chlorine, atrazine

Little information exists on the relationship between changes in biological structure in ecosystems and changes in functional processes. The ability of complex biotas to adapt to changing conditions ensures that vital system functions such as production, mineral cycling, and detrital processing are maintained through a broad range of environmental conditions. This has suggested to many ecosystem scientists that functional processes are inherently conservative and may not demonstrate adverse ecological effects at sufficiently low toxicant levels to warrant their

[1]Assistant professor, School of Forest Resources, Pennsylvania State University, University Park, PA 16802.
[2]Director, University Center for Environmental and Hazardous Materials Studies, and professor, Department of Biology, Virginia Polytechnic Institute and State University, Blacksburg, VA 24061.

inclusion in risk assessments. Alternatively, other scientists have used functional measures to detect small shifts in system function well before major shifts in taxonomic composition take place (for example, see Blanck [1]). Our ability to measure changes in taxonomic composition in relation to changing abiotic factors is well documented [2,3]; however, it is equally clear that shifts in organismal function (for example, see Lubinski et al. [4]) or community function [5] are sensitive to extremely low levels of perturbation.

A variety of organismal and community functional processes might be considered in developing risk assessment tools. In general, functional measurements are characterized by the measurement of changes in processes over time and are typically expressed as time-variant phenomena or processing rates: respiration rates, mineralization rates, production rates, and succession rates.

The authors of this paper hypothesized that toxic materials would adversely affect sensitive species, with the net result being the loss of species from communities. A loss of species should result in changes in colonization dynamics, production dynamics, nutrient assimilation, and other functional processes. We report here results of measurements of net dispersal rates of microbes in laboratory microcosms in response to toxic stress. Net dispersal rates to a new habitat can be measured as colonization rates, the rate of accrual of species on the new habitat. We have used microbial communities because they are easy to manipulate, can be maintained in the laboratory for extended periods, respond in the same range of toxic concentrations as higher taxa, and produce several generations within a short period of time. We have used measurements of colonization from complex communities as indices of toxic effects on complex biological systems. Effects can be measured as changes in colonization rate, as differences in the number of colonizing species at any given point in time, or as differences in community production.

The effective dispersal of propagules is a valuable functional attribute of complex communities since new habitats are routinely created by natural (and human-induced) disturbances. To maintain the complex interacting biota of any aquatic ecosystem requires that species be able to produce sufficient propagules to colonize the new habitat. The net rate of this process can be evaluated and compared among replicate communities exporting propagules to barren habitats while under toxic stress. We have also compared colonization dynamics and biomass accumulation (net production) in laboratory microcosms.

Methods

We conducted two types of tests to evaluate effects of toxic materials on microbial colonization: tests of effects on dispersal of protozoan species to artificial substrates and microcosm tests of effects on species dispersal and production. The dispersal tests were conducted using cadmium and a complex effluent containing copper and other heavy metals as toxicants. The microcosm tests examined the effects of chlorine (as hypochlorite) and atrazine on microbial communities.

The test systems used in these studies differ from other microcosm systems in several ways. A replicable microbial community is collected on a substrate of known heterogeneity. Sediment and associated surface-dwelling organisms (bacteria, algae, protozoa, and microinvertebrates) from a natural, unimpacted system are used as species sources in test systems. A barren habitat is provided on artificial substrates that are used as sampling devices. The test systems do not include large amounts of natural sediment (in contrast to the study by Giddings [6]), but they are amenable to continuous dosing with toxic material without the test community being washed out. The use of natural communities precludes the need to culture laboratory populations (see for comparison the study by Taub [7]).

Test Systems

Two test systems were used in these experiments. The first system (used for dispersal tests) was a static system in which toxic material was introduced once at the beginning of each experiment. These test systems were polyethylene tubs into which test media had been placed. Initially barren artificial substrates for collecting the microorganisms were placed around the perimeter of each testing system and a natural microbial community from a small pond on a similar artificial substrate was placed in the center of the tank to serve as the species source (Fig. 1a). The artificial substrates used were polyurethane foam (PF) blocks. These artificial substrates have been previously shown to be colonized like habitat islands [8–10]. Colonization of the initially barren artificial substrates in each tank was observed for 21 to 28 days.

The second test system (used for microcosm tests) allowed continuous replacement of toxic test media from stock solutions of varying concentrations diluted by a dechlorinated tap water diluent. A schematic diagram of this test system is shown in Figs. 1b and 2. In this test system, the initially barren artificial substrates were placed in the outfall half of the test tank, and two artificial substrates colonized in a natural ecosystem were placed in the influent half of the test tank. This ensured that the source community received the incoming diluted toxicant and that propagules from these substrate flowed toward the initially barren substrates in the outfall area of the tank. Three holes were cut in the outfall end of each polyethylene test tank to regulate the test medium volume to approximately seven litres. The flow volume through the tanks was adjusted so that no less than five volume replacements occurred per day.

Both test systems were illuminated with daylight-equivalent fluorescent lights (Vita-Lites, Durotest, Inc.) at an intensity of approximately 5000 lux. The photoperiods were 16 h light:8 h dark in the static tests and 12 h light:12 h dark in the continuous flow tests. The artificial substrates were attached to the bottoms of the test tanks by using small plastic hooks and silicon adhesive. Each substrate was attached to a hook by a loop of cotton or monofilament nylon line.

Sampling involved removing an initially barren artificial substrate from each test tank and immediately squeezing the substrate into a clean collecting vessel (a sterile polystyrene specimen cup) to remove the contents. This sampling technique does not remove all organisms; certain securely attached taxa may remain on the substrate after harvesting, but these taxa are a minor component of the total biomass harvested. The sample was allowed to settle prior to examination of the colonizing taxa. Settling usually required about 30 min. For several tests, subsamples were removed for determination of biomass parameters immediately after the substrates had been squeezed. To ensure that direct effects on the source community were not the reason for changes in the colonization of the initially barren, artificially barren substrates, the species source in each tank was examined at the conclusion of each experiment. Previous experiments

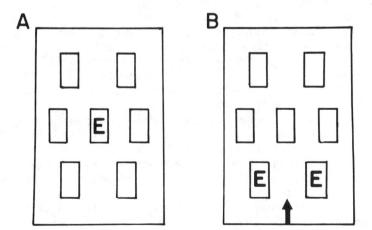

FIG. 1—*Test tanks for microbial colonization experiments:* (a) *arrangement of artificial substrates for the dispersal experiments;* (b) *arrangement of artificial substrates in the microcosms. E indicates naturally colonized species sources (epicenters).*

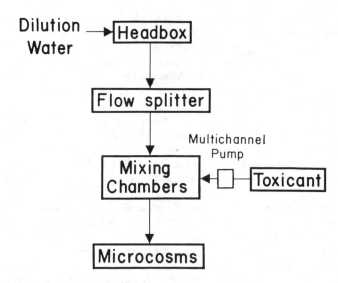

FIG. 2—*Schematic diagram of a microcosm testing system showing the route of continuous supply of diluent water and toxicant to the microcosm test tanks.*

have shown that, for low levels of toxicants, direct effects on the source communities are unlikely [11]. In addition, at the start of each experiment, three reference PF substrates were examined to provide an estimate of the number of taxa on the species sources, and these data were used in a comparison with the richness of the species source at the conclusion of each experiment.

Test Organisms

The PF substrates are general collectors of attached and surface-dwelling microbial species, including bacteria, algae, protozoa, and small metazoa. The most vagile members of these communities are the protozoa, whose species richness in most natural habitats ranges from 50 to 100 taxa. We examined the dispersal of this subset of the total microbial community, although other taxa were dispersing from the species sources to the initially barren substrates as well. Changes in biomass on the substrates over time were evaluated to make a gross measure of the influence of nonprotozoan taxa in the test systems. The chlorophyll *a* and adenosine triphosphate (ATP) biomass present per unit volume of sample were determined according to standard methods [12]. The protein biomass per unit volume of sample was estimated using a dye-binding, spectrophotometric method [13,14].

The numbers and kinds of protozoan species were determined by microscopically examining two or more subsamples from each substrate on each sampling day. Each subsample was systematically scanned, and the taxa were identified to the lowest practicable taxonomic unit, frequently to species. Where identification to the species level proved impossible, a note or a drawing was made to ensure continuity or identifications throughout the experiment. Cytological staining techniques were not used.

Source microbial communities were obtained by colonizing replicate PF substrates in a natural ecosystem for a period of 14 to 21 days. Previous experiments had shown that this period was sufficient to attain species equilibrium on the artificial substrates. The substrates were then collected *en masse* in an ice chest or clean plastic bucket partially filled with pond water. For the experiments reported here, microbial communities were obtained by colonizing artificial substrates in Pandapas Pond (Jefferson National Forest, Montgomery County, VA). The substrates were immediately returned to the laboratory and randomly allocated to already established test systems.

Toxicants

Dispersal experiments were conducted using test media (granulated wood charcoal filtered tap water) dosed with cadmium chloride (CdCl$_2$) (in the form CdCl$_2 \cdot 21/2$H$_2$O) and with a complex effluent containing copper. The mean measured cadmium doses were 0.4, 1.4, 2.7, 5.6, and 9.5 μg/L plus controls. The complex effluent was found to have an acid-extractable copper concentration of 98 μg/L. Effluent dilutions resulted in mean measured copper concentrations of 2.6 (control) 4.0, 6.6, 8.7, 12.2, and 18.0 μg/L.

The microcosm experiments used continuous toxicant dosing with chlorine (as hypochlorite) and atrazine (2-chloro-4-ethylamino)-6-isopropylamino-s-triazine). The mean measured doses of total residual chlorine (TRC) were 0, 2, 6, 24, 100, and 308 μg TRC/L. The measured atrazine doses were 3.1, 10, 32, 110, and 337 μg/L plus dilution water and carrier solvent (methanol) controls.

Experimental Design and Analysis

The experiments were conducted using triplicate test systems for each of five test concentrations plus controls. The concentrations of diluted toxicant ranged over two orders of magnitude. The concentrations for pure compound tests were determined based on literature reports of chronic toxicity. For complex effluents, the highest test concentration was 100%, and a "30%" (one-half order of magnitude) dilution series was used to achieve lower test concentrations.

The number of species established on artificial substrates over time was fitted to the MacArthur-Wilson [15] equilibrium model for island colonization

$$S = S_{eq}(1 - e^{-Gt})$$

This saturation model can be used to estimate the number of species at equilibrium (S_{eq}) and the rate at which colonization occurs (G) by fitting the number of species (S) sampled at differing times (t) to the model using nonlinear least squares regression methods of the Statistical Analysis System [16]. Estimates of the equilibrium species number or colonization rate can then be compared among test concentrations to determine the effects of toxic materials on the colonization process in microbial communities. Alternatively, differences among the test concentrations can be detected by a day-by-day basis using analysis of variance procedures [17]. We examined the fit of data to the equilibrium model using dummy variable analysis [18]. Regressions were also tested for significance and lack of fit, and an estimate of the amount of variation in the species-time data explained by the model equation (r^2) was obtained. Daily differences in the species numbers were detected using multiple comparisons with controls.

Results

Dispersal Tests

Dispersal of the protozoan species to habitat islands was severely affected by cadmium (Fig. 3). Reductions in the equilibrium species number were strongly dose related ($r^2 = 0.94$, $P < 0.01$). Estimates of the equilibrium species number were more informative than estimates of the colonization rate function (G), which did not show significant variation with dose, despite severe depressions in the number of colonizing species (Table 1). Biomass accrual on initially barren substrates also showed a dose-dependent response. The chlorophyll biomass (a measure

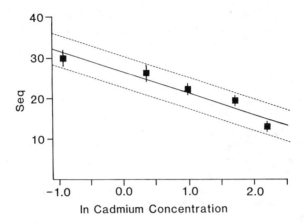

FIG. 3—*Relationship between the estimated equilibrium species number (S_{eq}) for protozoa and the cadmium dose in the dispersal experiments. The solid line is the best fit line from linear least squares regression ($r^2 = 0.94$, P < 0.01). The dashed lines indicate the 95% confidence interval around the regression line.*

TABLE 1—*Effect of cadmium on the protozoan colonization, chlorophyll a, and total biomass [adenosine triphosphate (ATP) biomass] in the dispersal experiments.*[a]

Cadmium, µg/L	S_{eq} (ASE)[b,c]	G (ASE)	Chlorophyll a, mg/L	ATP Biomass, mg/L[c]
Control	33.9 (2.1)	0.18 (0.04)	0.016	1.16
0.4	30.1 (2.0)	0.20 (0.05)	0.017	1.03
1.4	26.6* (1.9)	0.22 (0.06)	0.010	1.43
2.7	22.6* (1.8)	0.27 (0.09)	0.010	3.27*
5.6	19.4* (1.7)	0.32 (0.12)	0.009	2.45*
9.5	13.1* (1.6)	0.36 (0.20)	0.004*	3.61*

[a]S_{eq} = the estimated equilibrium species number; G = the colonization rate.
[b]The values in parentheses are asymptotic standard errors (ASE).
[c]The table values marked with an asterisk are significantly different from control values ($P < 0.05$).

of net algal production) was depressed by increasing cadmium concentration. However, the total biomass accumulation increased with cadmium concentration, probably because of the growth of metal-tolerant fungi. Elimination of algal biomass was also reflected in changes in the composition of protozoan functional groups [*19*] over the colonization period. Increased cadmium concentrations reduced the numbers of photosynthetic taxa (Fig. 4).

We estimated effective concentrations (ECs) based on proportional reductions in species numbers relative to those of controls. The EC_{20} (corresponding to a 20% reduction in the num-

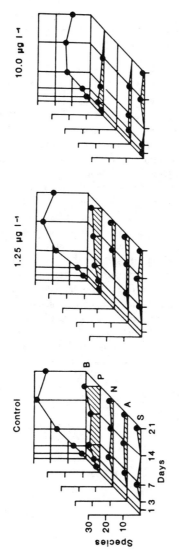

FIG. 4—*Effect of cadmium dose on the functional group composition of colonizing protoza. Key: B, bacterivores; P, photosynthetic protozoa; N, omnivores; S, saprotrophs; A, algivores; R, raptors.*

ber of species in comparison with control numbers) was 1.4 μg Cd/L. The EC_{05} (5% reduction in the species number), a value comparable to values derived for numerical water quality criteria [20], was 0.4 μg Cd/L.

Experiments using a complex effluent containing copper produced similar results (Fig. 5). Low effluent concentrations stimulated species colonization prior to eventual depression of species numbers. The highest test concentration (20% or ~ 18 μg Cu/L) significantly reduced the species number (by over 20%) relative to that of controls. This level of copper is near the current water quality criterion for copper [21]. There was evidence of stimulation of the species number in lower effluent (lower copper) concentrations. Other toxicants (including lead and chromium) were present at very low levels.

Microcosm Tests

Communities in the microcosm experiments showed a strong response to chlorine addition. The protozoan species number (estimated equilibrium species number) decreased with increasing chlorine dose and differed significantly from that of controls at treatment levels of 6 μg TRC/L and higher (Fig. 6). Only treatments with average measured chlorine doses of 2 μg/L produced results that did not differ significantly from those of controls. There was no significant effect on colonization rate (G) attributable to chlorine. There was some evidence of stimulation of production by low levels of chlorine after 28 days of colonization, and the biomass accumulation was nonlinearly related to the dose. The chlorophyll biomass was reduced 2 μg Cl/L but was elevated at 6 μg/L (Fig. 7a). The total biomass accumulation (Fig. 7b), as measured by the total extractable protein, was elevated at low chlorine doses and was not significantly lower than that of controls until the treatment concentration was 100 μg Cl/L, nearly ten times the current water quality criterion for chlorine (11 μg/L) [22].

Atrazine also produced stimulation at low doses in microcosms. The number of protozoan species (estimated equilibrium species number) was enhanced at atrazine levels between 3 and 30 μg/L (Fig. 8). The species numbers returned to control levels at 100 μg/L and were significantly reduced at 300 μg/L. These responses were reflected in the net production differences, and there were no significant effects attributable to the carrier solvent (methanol) used in the experiments. The chlorophyll biomass was enhanced at 3 and 10 μg atrazine/L and significantly depressed at 300 μg/L (Fig. 9a). The total protein biomass was significantly elevated at concen-

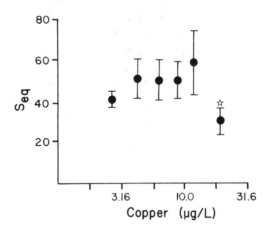

FIG. 5—*Relationship between the estimated equilibrium species number (S_{eq}) for protozoa and the copper dose in the dispersal experiments. The point marked with a star is significantly different from the control (P < 0.05).*

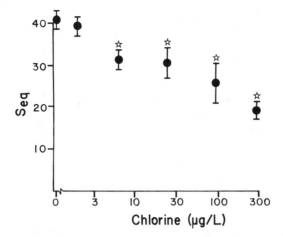

FIG. 6—*Relationship between the estimated equilibrium species number (S$_{eq}$) for protozoa and the chlorine dose in the microcosm experiments. Stars indicate significant differences from controls.*

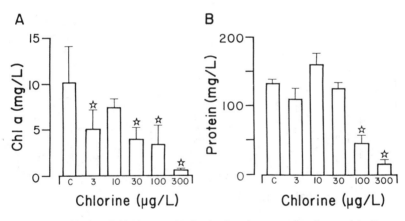

FIG. 7—*Effect of chlorine on biomass production in the microcosm experiments:* (a) *effect on chlorophyll concentration, a measure of algal biomass;* (b) *effect on protein concentration, a measure of total biomass. Stars indicate significant differences from controls (*P < 0.05).

trations from 3 to 30 μg/L and significantly reduced at 300 μg/L (Fig. 9b). The 100 μg/L treatment did not produce significantly different results from controls.

Discussion

Species dispersal and community development integrate the many environmental influences that affect reproduction. Toxic effects can be seen as a loss of species from a community and as a failure of species to disperse to a new habitat.

Dispersal Tests

The MacArthur-Wilson [15] theory of island colonization is essentially a birth-death-diffusion model of dispersal. As such, toxic effects might influence any or all of these processes. Our

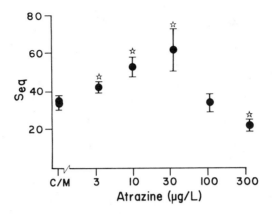

FIG. 8—*Relationship between the estimated equilibrium species number (S_{eq}) for protozoa and the atrazine dose in the microcosm experiments. C = control; M = methanol (carrier solvent) control.*

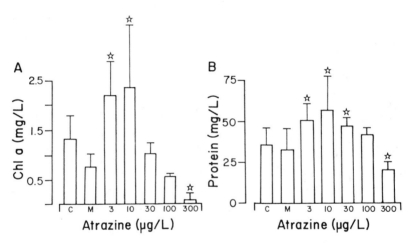

FIG. 9—*Effect of atrazine on biomass production in the microcosm experiments:* (a) *effect on chlorophyll concentration, a measure of algal biomass;* (b) *effect on protein concentration, a measure of total biomass. Stars indicate significant differences from controls. C = control; M = methanol (carrier solvent) control.*

observations of species dispersal in static testing system used to evaluate cadmium effects indicate that the number of species arriving on the habitat islands (artificial substrates) is similar for all treatments. That is, the total number of species observed in each test system was similar for all treatments, but the number of extant species detected at a given sampling was reduced by the toxic material. This suggests that cadmium affects population growth and survival of the migrating propagules. However, these effects might be either direct (for example, depression of the reproduction rate) or indirect (such as effects on the food supply). Diverse, interacting populations in communities might exhibit both effects.

We hypothesized that toxic materials would adversely affect sensitive species, with the net result being a loss of species from communities. Similar predictions have been made by Odum [23] for community succession under toxic stress. In tests with a copper-containing effluent, low copper concentrations were associated with apparent stimulation of species numbers. If species abundances are log-normally distributed [24,25], our observation of increased species numbers is probably a result of increases in abundance of species that were formerly undetectable by our sampling methods. This suggests that species survival or reproduction is positively affected by low levels of toxic materials. We discounted assumptions of nutrient effects, since we would have expected nutrients in the effluent to have resulted in increasing stimulation with increasing effluent concentration. The observed response may have been reflective of the subsidy-stress gradient postulated by Odum and his colleagues [26], although we would generally think of subsidy responses as being increases in production or processing rates and not increases in species richness.

Previous studies of chronic toxic effects of the metals used in dispersal tests agree with our estimates of adverse effects. Chronic single-species bioassays have shown adverse effects at concentrations between 0.08 and 11 μg Cd/L [20]. Tests of effects on microcosms have produced estimates of adverse impact at concentrations ranging from 0.2 to 10 μg Cd/L [27,28]. Our estimates of adverse effect ($EC_{20} \simeq 1$ μg Cd/L) agree with present water quality criteria. Similarly, reduction of the species number in systems dosed with a copper-containing effluent occurred at levels comparable to current water quality criteria [21] at the test water hardness. The numerical criterion is 14 μg Cu/L at a water hardness of 65 mg calcium carbonate ($CaCO_3$)/L; species numbers also decreased significantly at copper concentrations of 18 μg/L.

Microcosm Tests

The relationship between the number of protozoan species and the net production in surface (periphyton) communities in response to toxic stress was more apparent in the microcosm tests. As in the dispersal tests, some toxic inputs stimulated the species number. This stimulation was also reflected in differences in net biomass accumulation in microcosms.

Significant reductions in the species number occurred at lower levels of chlorine (2 to 6 μg/L) than the current criterion level of 11 μg/L [22], which is based on only three studies of chronic chlorine toxicity. However, the biomass production did not demonstrate the toxic effect of chlorine at these low levels.

Atrazine is not considered toxic to animals but may produce secondary or indirect effects in aquatic systems when primary production is reduced. High atrazine concentrations (>1900 μg/L) are necessary to detect toxicity to fish (see Ward and Ballantine [29]). In microcosms and mesocosms, adverse effects have been observed at >300 μg/L [30], but effects (chronic EC_{50}, or 50% reduction in the species number) on microcosm and mesocosm components have occurred at concentrations ranging from 82 to 165 μg/L for effects on the chlorophyll biomass and oxygen production [31]. We observed significant enhancement of response at atrazine levels below 30 μg/L and did not observe reduced responses until atrazine concentrations exceeded 110 μg/L.

Nonlinear responses of communities to low levels of toxic stress provide evidence of both disruption of normal community functioning (as exemplified by controls) and accommodation of communities to toxic stress at certain low levels of stress. For example, we found concomitant increases in the species number and biomass in systems exposed to chlorine and atrazine stress. Both positive and negative deviations of communities from nominal (control) states may be evidence of toxic stress. While the dynamics of these stimulations responses resemble the hypothesized subsidy-stress gradient [26], the "subsidy" is sometimes difficult to identify. Chlorine is a powerful oxidant and could provide nutrient material to communities by oxidizing organic material. However, naked species such as protoza should also be readily oxidized. The mechanism that would produce increased species numbers and biomass from this stress is not apparent.

Atrazine affects photosynthesis by uncoupling the flow of electrons leading to photophosphorylation. In microcosms, it stimulated the production of chlorophyll biomass at low doses and resulted in an increase in the protozoan species number. As in other experiments, the species number and biomass showed similar responses to a toxic dose. Since atrazine is not readily degraded by microbial communities, it probably stimulates production by some unknown mechanism. We speculated, for example, that atrazine addition may increase electron flow during photosynthesis. This may temporarily augment primary production. However, it is difficult to interpret a known toxic addition as a subsidy when its effects cause large changes in relatively gross measurements of community structure.

The subsidy effect might also result from interference with normal interactions in communities. Whether the observed stimulation results from the loss of keystone species, species substitution, alteration of normal succession, or species turnover is not known. However, it appears that functional processes such as net production are reflected by gross changes in microbial community structure. This substantiates the close linkage of structure and function. Further, positive deviation may also reflect "adverse" changes in communities. While these changes may differ from our common expectation of depression of the species number and processing rate, they may signal important disruptions in the community dynamics produced by toxic materials.

More detailed investigations of the specific taxonomic shifts that occur as a result of toxic stress might further explain the observed patterns. However, protozoan communities may not be ideal candidates for such investigations. Species turnover rates are extremely high [32] and stochastic differences between adjacent communities may preclude detecting important shifts in key species. Protozoan communities are trophically diverse and may be more representative of the diversity of real ecosystems than taxonomically more restricted communities such as attached algae.

Communities can integrate toxic effects and display effects on emergent functional properties of systems that cannot be simulated in single-species tests. The testing systems described in this paper can be useful in examining these effects and can be developed to investigate site-specific situations. The use of receiving system communities to examine effects may provide information that can be more easily validated than predictions from standard bioassays. However, caution should be exercised in evaluating effects on stream communities that may have already developed tolerances to toxicants from upstream sources. A variety of community functional measurements might provide additional information upon which to base judgements of environmental safety and harm. These measurements might include activities of selected enzyme groups, uptake of vital nutrients, and other measures that would complement investigations of changes in the species composition and biomass production. Such investigations would add measurably to our understanding of the relationship between community structure and function in response to toxic stress.

Acknowledgments

This research was supported, in part, by grants from the U.S. Air Force Office of Scientific Research (Grant No. AFOSR-85-0324), the U.S. Environmental Protection Agency (Grant No. R-812813-01-1), the E. I. duPont de Nemours Education Foundation, and Environmental and Water Resources Research (EWRR). (This article has not been subjected to the U.S. Environmental Protection Agency's peer and policy review; it does not necessarily reflect the views of that agency, and no official endorsement should be inferred.) Technical assistance was provided by B. R. Niederlehner and N. J. Bowers. The editorial assistance of Darla Donald in preparing this chapter for publication is gratefully acknowledged.

References

[1] Blanck, H., *Hydrobiologia,* Vol. 124, 1985, pp. 251-261.
[2] Gauch, H., "ORDIFLEX—A Flexible Computer Program for Four Ordination Techniques: Weighted Averages, Polar Ordination, Principal Component Analysis, and Recriprocal Averaging, Release B," *Ecology and Systematics,* Cornell University, Ithaca, NY, 1977.

[3] Pielou, E., *Interpretation of Ecological Data: A Primer on Classification and Ordination*, Wiley, New York, 1984.
[4] Lubinski, K. L., Cairns, J., Jr., and Dickson, K. L. in *Trace Substances in Environmental Health*, D. D. Hemphill, Ed., University of Missouri, Columbia, MO, 1978, pp. 508-514.
[5] Bobbie, R. J. and White, D. C., *Applied and Environmental Microbiology*, Vol. 39, 1980, pp. 1212-1222.
[6] Giddings, J. M. in *Hazard Assessment of Chemicals—Current Developments*, Vol. 2, J. Saxena, Ed., Academic Press, New York, 1983, pp. 45-94.
[7] Taub, F., *International Journal of Environmental Studies*, Vol. 10, 1976, pp. 23-33.
[8] Cairns, J., Jr., Dahlberg, M. L., Dickson, K. L., Smith, N., and Waller, W. T., *American Naturalist*, Vol. 103, 1969, pp. 439-454.
[9] Cairns, J., Jr., Kuhn, D. L., and Plafkin, J. L. in *Methods and Measurements of Periphyton Communities: A Review, ASTM STP 690*, R. L. Weitzel, Ed., American Society for Testing and Materials, Philadelphia, 1979, pp. 34-57.
[10] Bamforth, S. S. in *Artificial Substrates*, J. Cairns, Jr., Ed., Ann Arbor Science Publishers, Ann Arbor, MI, 1982, pp. 115-130.
[11] Niederlehner, B. R., Pratt, J. R., Buikema, A. L., Jr., and Cairns, J., Jr., *Environmental Toxicology and Chemistry*, Vol. 4, 1985, pp. 155-165.
[12] American Public Health Association, American Water Works Association, and Water Pollution Control Federation, *Standard Methods for the Examination of Water and Wastewater*, 16th ed., Washington, DC, 1985.
[13] Bradford, M. M., *Analytical Biochemistry*, Vol. 72, 1976, pp. 248-254.
[14] Rausch, P., *Hydrobiologia*, Vol. 78, 1981, pp. 237-251.
[15] McArthur, R. H. and Wilson, E. O., *The Theory of Island Biogeography*, Princeton University Press, Princeton, NJ, 1967.
[16] *SAS User's Guide: Statistics*, SAS Institute, Cary, NC, 1982.
[17] Sokal, R. R. and Rohlf, F. J., *Biometry*, 2nd ed., W. H. Freeman, San Francisco, 1983.
[18] Kleinbaum, D. and Kupper, L., *Applied Regression Analysis and Other Multivariate Methods*, Duxbury Press, North Scituate, MA, 1978.
[19] Pratt, J. R. and Cairns, J., Jr., *Journal of Protozoology*, Vol. 32, 1985, pp. 409-417.
[20] U.S. Environmental Protection Agency, "Ambient Water Quality Criteria for Cadmium," EPA-440/5-80-025, National Technical Information Service, Springfield, VA, 1980.
[21] U.S. Environmental Protection Agency, "Ambient Water Quality Criteria for Copper," EPA-440/5-80-036, National Technical Information Service, Springfield, VA, 1980.
[22] U.S. Environmental Protection Agency, "Ambient Water Quality Criteria for Chlorine," EPA-440/5-84-030, National Technical Information Service, Springfield, VA, 1984.
[23] Odum, E. P. in *Stress Effects on Natural Ecosystems*, G. Barret and R. Rosenberg, Eds., Wiley, New York, 1981.
[24] Preston, F. W., *Ecology*, Vol. 29, 1948, pp. 254-283.
[25] Patrick, R., *Proceedings of the National Academy of Sciences*, Vol. 58, 1967, pp. 1335-1342.
[26] Odum, E. P., Finn, J. T., and Franz, E. H., *BioScience*, Vol. 29, 1979, pp. 349-352.
[27] Marshall, J. S. and Mellinger, D. L., *Canadian Journal of Fisheries and Aquatic Sciences*, Vol. 37, 1980, pp. 403-414.
[28] Heath, R. T., *International Journal of Environmental Studies*, Vol. 13, 1979, pp. 87-93.
[29] Ward, F. S. and Ballantine, L., *Estuaries*, Vol. 8, 1985, pp. 22-27.
[30] Stay, F. S., Larsen, D. P., Katko, A., and Rohm, C. M. in *Validation and Prediction of Laboratory Methods for Assessing the Fate and Effects of Contaminants in Aquatic Ecosystems, ASTM STP 865*, T. P. Boyle, Ed., American Society for Testing and Materials, Philadelphia, 1985, pp. 75-90.
[31] Larsen, D. P., DeNoyelles, F., Jr., Stay, F., and Shiroyama, T., *Environmental Toxicology and Chemistry*, Vol. 5, 1986, pp. 179-190.
[32] Schoener, T., *Oikos*, Vol. 41, 1983, pp. 372-377.

Philippe Ross,[1] *Vincent Jarry,*[2] *and Harm Sloterdijk*[3]

A Rapid Bioassay Using the Green Alga *Selenastrum capricornutum* to Screen for Toxicity in St. Lawrence River Sediment Elutriates

REFERENCE: Ross, P., Jarry, V., and Sloterdijk, H., **"A Rapid Bioassay Using the Green Alga** ***Selenastrum capricornutum*** **to Screen for Toxicity in St. Lawrence River Sediment Elutriates,"** *Functional Testing of Aquatic Biota for Estimating Hazards of Chemicals, ASTM STP 988,* J. Cairns, Jr., and J. R. Pratt, Eds., American Society for Testing and Materials, Philadelphia, 1988, pp. 68–73.

ABSTRACT: A toxicity bioassay using the green alga *Selenastrum capricornutum* and measuring inhibition of photosynthetic carbon-14 (^{14}C)-labeled carbon dioxide ($^{14}CO_2$) assimilation is described. The essential test specifications are the following:

(*a*) experimental vessels: 16-mL test tubes;
(*b*) exposure and incubation times: 20 h exposure plus 4 h incubation;
(*c*) dilution water: demineralized H_2O;
(*d*) algal cell density: 1.9×10^5 cells mL^{-1};
(*e*) radioactivity: 0.072 μCi mL^{-1}, as ^{14}C-labeled sodium bicarbonate [$Na_2(H^{14}CO_3)_2$];
(*f*) nutrient enrichment: 1 mL of $\times 10$ provisional algal assay procedure (PAAP) medium; and
(*g*) isotope partitioning: acidification and bubbling.

The resulting method is rapid, sensitive, and reliable (coefficient of variation $< 10\%$) and is applied to elutriates of St. Lawrence River sediments in an ongoing research project.

KEY WORDS: hazard evaluation, toxicity, bioassays, algae, *Selenastrum capricornutum*, sediments, elutriates, carbon-14-labeled carbon dioxide ($^{14}CO_2$), $^{14}CO_2$ uptake, St. Lawrence River, copper chloride

The green alga *Selenastrum capricornutum* Printz (Chlorophyta; Chlorophyceae; Chlorococcales) became a widely used bioassay organism with publication of the provisional algal assay procedure (PAAP) [1] and the algal assay procedure bottle test (AAPBT) [2,3]. Early experiments with this alga were oriented towards measurement of the effects of nutrient additions [4-11], but its value for toxicity assessment was soon recognized [12-32], and *S. capricornutum* has now become a standard tool in aquatic ecotoxicology. Most researchers have retained the use of batch cultures in bottles as their experimental units, but some recent efforts at miniaturization have appeared [33,34].

The standard bioassay measures the growth of the culture through direct counts of cell num-

[1]Associate aquatic toxicologist, Illinois Natural History Survey, Champaign, Il 61820.
[2]Agent de recherche, Département de Sciences biologiques, Université de Montréal, Montréal, Québec, Canada H3C 3J7.
[3]Survey scientist, Inland Waters Directorate, Environment Canada, Longueuil, Québec, Canada J4K 1A1.

bers [*16,22*], automated particle counting [*19,21,23,24,28,31–34*], or dry-weight biomass estimates [*8*]. The incubation periods are usually 4 [*19,32,33*], 7 [*10,16,19*], or 8 days [*23,31,33,34*]. *Selenastrum capricornutum* has also been used in bioassays measuring inhibition of dichlorophenyldimethylurea (DCMU)-induced fluorescence (a 24-h test) [*24*], oxygen evolution (1-h procedure) [*27*], and cellular adenosine triphosphate (ATP) (4 h or 4, 7, or 8 days) [*24,29,33*].

The use of inhibition of photosynthetic uptake of carbon-14 (^{14}C)-labeled carbon dioxide ($^{14}CO_2$) as the measure of toxicity in *Selenastrum* tests is a relatively recent development. Such tests have previously used 20 or 24-h toxicant exposure periods in 100 [*30*], 125 [*24*], or 500-mL [*26*] flasks. The exposure has then been followed by incubation with the radioisotope for periods of 10 min [*24*], 2 h [*26*], or 4 h [*30*].

The present report is the result of three years of effort to develop a rapid, cost-effective algal bioassay in relatively small vessels for use in screening sediment elutriates for toxic potential. Special attention has been paid to reducing the variability of the test to an acceptable level.

The Study Area

In 1984 the authors of this paper began a collaborative project (Environmental Canada and l'Université de Montréal) to study the hazard associated with contamination of sediments in the depositional basins of the St. Lawrence River system in the vicinity of Montréal, Canada. These basins receive dissolved and suspended loads of contaminants from the Great Lakes–St. Lawrence River Corridor and the Ottawa River basin, as well as locally generated pollutants [*35*]. Systematic sampling of the depositional basins began in May 1984 and was followed by chemical analysis of the sediments [*36*] and their elutriates [*37*] and by mapping of the contaminant distributions [*36*]. A battery of bioassays was performed on each elutriate.

The Basic Method

The bioassays involve the calculation of a photosynthesis inhibition curve for each elutriate. The algal culture (*Selenastrum capricornutum* Printz, obtained from the Environmental Protection Service, Longueuil, Québec) is incubated at 26 ± 3°C with a 12 h light:12 h dark photoperiod in PAAP medium [*38*] for 6 days before each bioassay. At the beginning of the test, culture is added to the experimental vessels at a density of 1.9×10^5 cells mL^{-1}. After an adaptation period, the control and treatment vessels receive inoculations of carbon-14 as ^{14}C-labeled sodium bicarbonate [$Na_2(H^{14}CO_3)_2$]. A series of at least six concentrations of elutriate is used for each test, and the results are expressed as percentages of the control photosynthesis.

Developmental Objectives

We began with a standard ecological technique for measuring algal photosynthesis by $^{14}CO_2$ uptake [*39–41*]. We converted this to a useful toxicity bioassay by accomplishing the following objectives: smaller incubator space requirements, reduced simple volumes, lower cost of materials, greater sensitivity to contaminants, and increased statistical reliability.

Space, Sample Volume, and Materials

By changing our incubation vessels from 128-mL bottles to 16-mL test tubes, we eliminated our incubator space problem and greatly reduced the volume of elutriate needed for the bioassays. We are now also able to use small quantities of isotope, adding only 1.15 μCi to each tube, while actually increasing the available isotope per millilitre of culture volume to 0.072 μCi mL^{-1}.

Sensitivity

The lack of response in some of our preliminary tests led us to suspect that stimulation, caused by enrichment from dissolved nutrients in the elutriates, could be masking the toxic (inhibitory) effects. Parallel tests using known concentrations of a single toxicant [copper chloride ($CuCl_2$)] showed low toxicity and high variability in the unenriched tubes (Fig. 1a) in comparisons with a similar test with PAAP algal culture medium [38] added (Fig. 1b). By adding enough nutrient to preclude growth limitation at all exposure concentrations, the toxic responses (inhibition relative to controls) were isolated [24] and the sensitivity of the tests as a toxicity bioassay was improved. Consequently, we now add 1 mL of concentrated ($\times$10) PAAP algal culture medium to each tube.

The period of exposure to the toxic material (adaptation period) before incubation with the radioisotope was also found to be critical. Again using known concentrations of $CuCl_2$, a series of tests was performed with total exposure periods of 4 h (no adaptation phase), 24 h (20-h adaptation), and 48 h (44-h adaptation). The 24-h tests were much more sensitive than the 4-h tests, but a 48-h period yielded no further improvement.

In order to apply the modified test to our study material, St. Lawrence River sediment elutriates, the question of dilution water still had to be considered. Surface waters, even though filtered, may contain dissolved substances capable of complexing or chelating toxicants, thereby altering their observed toxicity [42,43]. In another series of tests we compared the toxicity of elutriates using two diluents, St. Lawrence River water and demineralized water. The toxicity was much higher, and the variability much lower, with demineralized water (Fig. 2).

Reliability

We found filtration to be an unacceptable method of separating assimilated from unassimilated isotope. Filtration frequently took longer than 10 min, which created inefficient backlogs in the sample processing flow. Filtration times are highly variable, which may have been the cause of the elevated coefficients of variation, which were as high as 25%. We now use the acidification-bubbling method to drive off unassimilated radiocarbon [44,45]. After the 4-h incubation, a 3-mL subsample is removed from each culture tube and added to a scintillation vial. The subsample is then acidified with 100 μL of 0.1 N hydrochloric acid (HCl) and bubbled with air for 5 min. Ten millilitres of a two-phase scintillation cocktail (Aquasol-2, New England Nuclear) is added to each vial, followed by counting on an LKB-Wallac Rack-Beta counter with automatic quench correction. This procedure obviates the need for filtration, saving time and dramatically reducing the variability between replicates. The coefficient of variation for our bioassays is now always less than 10%.

Conclusions

There are several advantages to a photosynthesis bioassay. The test takes only 24 h to complete. Smaller vessels, and therefore less space, can be used. Finally, experimental units can be easily replicated to provide more degrees of freedom where needed. We are now satisfied that we have a sensitive, reliable bioassay that performs well with single toxicants and equally well with complex unknown toxicant mixtures (for example, sediment elutriates). Results of this bioassay on elutriates from St. Lawrence River sediments indicate that sediments from some stations are toxic while others are not. A statistical premodel, using correlation and multiple regression

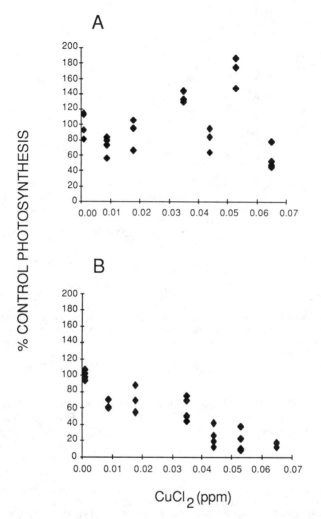

FIG. 1—*Photosynthetic response of* Selenastrum capricornutum *to varying concentrations of CuCl₂:* (a) *4-h exposure with no nutrient enrichment;* (b) *4-h exposure with nutrient enrichment with PAAP medium.*

techniques, showed a significant inhibitory effect by arsenic in the elutriates from the various stations [46].

Acknowledgments

This research was supported by a "Toxfund" grant from Environmental Canada (to H. Sloterdijk) and a Strategic Grant from the Natural Sciences and Engineering Research Council (NSERC) of Canada (Grant No. G1571, to P. Ross et al.) The help of personnel from the Inland Waters Directorate laboratories at Longueuil, Québec, and Burlington, Ontario, was provided through their respective operating budgets. P. Ross was an NSERC university research fellow and V. Jarry was employed through the Programme d'Emplois d'Eté Axés sur la Carrière (PEEAC). We are grateful for the help of summer students working on the project, including

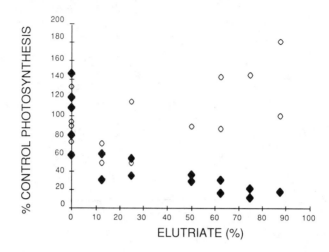

FIG. 2—*Photosynthetic response of* Selenastrum capricornutum *to varying concentrations of Lake St. Pierre sediment elutriate when treated with exposure and nutrient enrichment with PAAP medium:* (open symbols) *tests with filtered St. Lawrence River water as the diluent:* (closed symbols) *tests with demineralized water as the diluent.*

F. Blanchette and B. Laplante (NSERC summer research awards), D. Léger and J. Parent (Manpower and Immigration awards), and J.-F. Doyon and N. Rondeau. Computer time was furnished by the Centre de Calcul at l'Université de Montréal.

References

[1] U.S. Environmental Protection Agency, "Provisional Algal Assay Procedure," Joint Industry/Government Task Force on Eutrophication, New York, NY, 1969.

[2] U.S. Environmental Protection Agency, "Algal Assay Procedure Bottle Test," National Eutrophication Research Program, U.S. Environmental Protection Agency, Corvallis, OR, 1971.

[3] Miller, W. E., Greene, J. C., and Shiroyama, T., "The *Selenastrum capricornutum* Printz Algal Assay Bottle Test: Experimental Design, Application, and Data Interpretation Protocol," Report No. EPA-600/9-78-018, U.S. Environmental Protection Agency, Corvallis, OR, 1970.

[4] Miller, W. E. and Maloney, T. E., *Journal of the Water Pollution Control Federation,* Vol. 43, 1971, pp. 2361-2365.

[5] Maloney, T. E., Miller, W. E., and Blind, N. L., in *Advances in Water Pollution Research,* Pergamon Press, New York, 1972, pp. 205-214.

[6] Chiaudani, G. and Vighi, M., *Water Research,* Vol. 8, 1974, pp. 1063-1069.

[7] Miller, W. E., Maloney, T. E., and Greene, J. C., *Water Research,* Vol. 8, 1974, pp. 667-679.

[8] Greene, J. C., Miller, W. E., Shiroyama, T., and Maloney, T. E., *Water, Air, and Soil Pollution,* Vol. 4, 1975, pp. 415-434.

[9] Payne, A. G., *Water Research,* Vol. 9, 1975, pp. 937-955.

[10] Shiroyama, T., Miller, W. E., and Greene, J. C., "Effect of Nitrogen and Phosphorus on the Growth of *Selenastrum capricornutum,*" *Proceedings,* Biostimulation and Nutrient Assessment Workshop, Report No. EPA-660/3-75-034, U.S. Environmental Protection Agency, Washington, DC, 1975, pp. 132-142.

[11] "Etude intégrée de la qualité des eaux des bassins versants des rivières Saint-François et Yamaska," Scientific Report No. 52, Vol. 2, *Secteur des substances nutritives,* Institut national de la recherche scientifique—Eau, 1976.

[12] Bartlett, L., Rabe, F., and Funk, W. H., *Water Research,* Vol. 8, 1974, pp. 179-185.

[13] Klotz, R. L., Cain, J. R., and Trainor, F. R., *Journal of Phycology*, Vol. 11, 1975, pp. 411-414.

[14] Maloney, T. E. and Miller, W. E. in *Water Quality Parameters, ASTM STP 573*, American Society for Testing and Materials, Philadelphia, 1975, pp. 344-355.

[15] Greene, J. C., Miller, W. E., and Shiroyama, T. in *Proceedings*, Symposium on Terrestrial and Aquatic Ecological Studies of the Northwest, Cheney, WA, 26-27 March 1976, pp. 327-335.

[16] Monahan, T. J., *Journal of Phycology*, Vol. 12, 1976, pp. 358-362.

[17] "Impact de flottage du bois sur les eaux du lac Talbot: évaluation à l'aide de tests biologiques," Scientific Report No. 77, Institut national de la recherche scientifique—Eau, 1977.

[18] Sachdev, D. R. and Clesceri, N. L., *Journal of the Water Pollution Control Federation*, Vol. 50, 1978, pp. 1810-1820.

[19] Chiaudani, G. and Vighi, M., *Mitteilungen, International Association of Theoretical and Applied Limnology*, Vol. 21, 1978, pp. 316-329.

[20] Hendricks, A. C., *Journal of the Water Pollution Control Federation*, Vol. 50, 1978, pp. 163-168.

[21] Christensen, E. R., Scherfig, J., and Dixon, P. S., *Water Research*, Vol. 13, 1979, pp. 79-92.

[22] Eloranta, V. and Eloranta, P., *Annales Botanici Fennici*, Vol. 17, 1980, pp. 26-34.

[23] Joubert, J., *Water Research*, Vol. 14, 1980, pp. 1759-1763.

[24] Couture, P., Van Coillie, R., Campbell, P. G. C., and Thellen, C., *Les Colloques de l'INSERM*, Vol. 106, 1981, pp. 255-272.

[25] Eloranta, V. and Laitinen, O., *Vatten*, Vol. 3, 1982, pp. 317-331.

[26] Laegrid, M., Alstad, J., Klaveness, D., and Selp, M., *Environmental Science and Technology*, Vol. 17, 1983, pp. 357-361.

[27] Bennet, P. H. and Sandmann, E., *Chemosphere*, Vol. 12, 1983, pp. 1047-1054.

[28] Allen, H. E., Blatchley, C., and Brisbin, T. D., *Bulletin of Environmental Contamination and Toxicology*, Vol. 30, 1983, pp. 448-455.

[29] Blaise, C., Legault, R., Bermingham, N., Van Coillie, R., and Vasseur, P., *Sciences et Techniques de l'Eau*, Vol. 17, 1984, pp. 245-250.

[30] Eloranta, V., Halttunen-Keyrilainen, L., and Kuivasniemi, K., *Sciences et Techniques de l'Eau*, Vol. 17, 1984, pp. 267-274.

[31] Blaise, C. and Couture, P., *Hydrobiologia*, Vol. 114, 1984, pp. 39-50.

[32] Kuivasniemi, K., Eloranta, V., and Knuutinen, J., *Archives of Enivronmental Contamination and Toxicology*, Vol. 14, 1985, pp. 43-49.

[33] Blaise, C., Legault, R., Bermingham, N., Van Coillie, R., and Vasseur, P., *Toxicity Assessment*, Vol. 1, 1986, pp. 261-281.

[34] Blaise, C., *Toxicity Assessment*, Vol. 1, 1986, pp. 377-385.

[35] Allan, R. J., *Water Pollution Research Journal of Canada*, Vol. 21, 1986, pp. 1-19.

[36] Jarry, V., Ross, P., Champoux, L., Sloterdijk, H., Mudroch, A., Couillard, Y., and Lavoie, F., *Water Pollution Research Journal of Canada*, Vol. 20, 1985, pp. 75-99.

[37] Champoux, L., Ross, P. E., Jarry, V., Sloterdijk, H., Mudroch, A., and Couillard, Y., *Revue Internationale des Sciences de l'Eau*, Vol. 2, 1986, pp. 95-107.

[38] Stein, J. R., *Culture Methods and Growth Measurements*, Cambridge University Press, Cambridge, England, 1973.

[39] Steeman-Nielsen, E., *Journal du Conseil*, Conseil International pour l'Exploration de la Mer, Vol. 18, 1952, pp. 117-140.

[40] Vollenweider, R. A., Munawar, M., and Stadelman, P., *Journal of the Fisheries Research Board of Canada*, Vol. 31, 1974, pp. 739-762.

[41] Fitzwater, S. E., Knauer, A. K., and Martin, J. H., *Limnology and Oceanography*, Vol. 27, 1982, pp. 544-551.

[42] *Water Quality Criteria*, National Academy of Sciences, Washington, DC, 1972.

[43] Wang, W., *Environmental Pollution (Series B)*, Vol. 11, 1986, pp. 193-204.

[44] Schindler, D. W., Schmidt, R. W., and Reid, R. A., *Journal of the Fisheries Research Board of Canada*, Vol. 29, 1972, pp. 1627-1631.

[45] Gächter, R., Mares, A., and Tilzer, M. M., *Journal of Plankton Research*, Vol. 7, 1984, pp. 359-364.

[46] Jarry, V., "Répartition spatiale et effet phytotoxique des contaminants dans les sédiments du lac St.-Louis (fleuve St.-Laurent)," M.Sc. thesis, Département de Sciences biologiques, Université de Montréal, Montreal, Canada, 1986.

Scott D. Bridgham,[1] *Donald C. McNaught,*[2] *and Craig Meadows*[2]

Effects of Complex Effluents on Photosynthesis in Lake Erie and Lake Huron

REFERENCE: Bridgham, S. D., McNaught, D. C., and Meadows, C., **"Effects of Complex Effluents on Photosynthesis in Lake Erie and Lake Huron,"** *Functional Testing of Aquatic Biota for Estimating Hazards of Chemicals, ASTM STP 988,* J. Cairns, Jr., and J. R. Pratt, Eds., American Society for Testing and Materials, Philadelphia, 1988, pp. 74–85.

ABSTRACT: Phytoplankton are the base of the food chain in most large lake ecosystems; if affected by environmental pollutants, significant ecosystem changes can result with potential impact on higher trophic levels. This research determined the effects of a complex effluent discharge from the River Raisin in Monroe County, Michigan, on the Lake Erie ecosystem. This river flows through southern Michigan and has large nutrient and industrial inputs, especially in the Monroe Harbor area. The functional parameters measured were bacterial uptake rate of acetate, zooplankton feeding and reproduction rates, and primary production. The results of the effects of complex effluents on gross photosynthesis, measured as carbon-14 (^{14}C) uptake, are presented in this paper.

Intensive sampling was undertaken during the ice-free seasons of 1983 and 1984. In 1983, water from various stations in the River Raisin was mixed with eutrophic Lake Erie water at several dilutions. The rate of photosynthesis in the mixed samples was compared with that of a control composed of Lake Erie water only. No clear trend in photosynthetic response was evident with increasing percentages of river station water. In 1984, oligotrophic Lake Huron water was substituted as the control, resulting in massive stimulation in photosynthesis with increasing percentages of river station water. There were increased levels of nutrients, toxicants, and algal biomass in the harbor area in relation to the Lake Huron control. The 1984 data were evaluated again after being weighted for the relative biomass of algae as ^{14}C uptake, so that a nutrient-toxicant interaction could be examined. Using these simple linear mixing criteria, a gradient in toxicity was found, with the upstream area of the river causing stimulation of photosynthesis while the harbor stations caused inhibition at all concentrations of effluent. Extensive chemical and physical analyses were done on the station effluents and controls. An analysis of covariance on the 1984 data indicated that no single toxicant was responsible for the inhibition of photosynthesis found in the harbor area.

Results suggest that, in complex effluent tests when an oligotrophic control is used with eutrophic effluents, a weighting factor for the relative biomass of algae must be used. Predictable trends occurred when this was done in relation to the Lake Huron control. Tributaries with diverse anthropogenic inputs, such as the River Raisin, have the potential for significant inhibition of primary production in receiving bodies of water.

KEY WORDS: complex effluents, photosynthesis, River Raisin, Great Lakes, nutrient-toxicant interactions, weighted algal biomass

A diverse array of pollutants can occur together in bodies of water. Nevertheless, single pollutants are normally tested for their effects on aquatic biota under carefully controlled laboratory

---

[1]Graduate student, School of Forestry and Environmental Studies, Duke University, Durham, NC 27706.

[2]Professor and technician, respectively, Department of Ecology and Behavioral Biology, University of Minnesota, Minneapolis, MN 55455.

conditions. The regulatory protocol in the United States has been, and probably will continue to be, based on these single-species, single-toxicant tests because of convention, economy, and the clear-cut interpretation of the results [1-3]. Such tests, however, are more of a comparative than a predictive tool [4]. Traditional limnological studies on photosynthesis have focused on the effects of either nutrients or toxicants on single species of algae or community assemblages, but rarely have the combined effects of nutrients and toxicants been examined.

Ecosystems are composed of complex assemblages of biotic and abiotic variables which define and regulate the system. Physical conditions are often quite variable spatially and temporally. Aquatic ecosystems are usually assaulted by many simultaneous anthropogenic perturbations, often with any single toxicant at a subthreshold level and with cultural eutrophication as a coincident problem. The sum effect of the array of pollutants with their synergisms and antagonisms, all against the background of the complicated milieu of the normal functioning of the ecosystem, is most often an unknown quantity. It cannot be assumed that single-species toxicity testing will adequately predict the responses of components within an ecosystem [5-7] or that laboratory microcosms effectively mimic actual ecosystem-level processes [1,4,8]. Much of the ecotoxicology research that has been done in situ, testing actual site-specific effects, has had a taxonomic (that is, structural) bias; however, changes in community structure are not necessarily related to ecosystem (that is, functional) changes [9]. Effects of toxicants on photosynthesis obtained under one nutrient regime may not be the same as those under a different nutrient regime. It has been theorized that an ecosystem is more than the sum of a random assemblage of species in a particular habitat, but rather an integrated functioning unit with significant ability for self-regulation and homeostasis [9-13]. Ecosystems are likely to possess both resilience and resistance to perturbation. The ultimate goal of environmental protection is the maintenance of the integrity of the ecosystem, including both its structural and functional components [9,14].

This paper presents the results from part of a multidisciplinary study that used complex effluent tests to examine the effects of the River Raisin, a heavily polluted river flowing through southeastern Michigan, on the Lake Erie ecosystem. Also examined were the effects of the River Raisin on a more oligotrophic alternative, using Lake Huron water as the control. Ecosystem processes were tested at three trophic levels—primary producers, herbivores, and decomposers—with each assay being a true systems level test [6]. The objectives of the broader study were the following: (1) to determine the effects of complex effluents of the River Raisin on the biota of the Great Lakes; (2) to determine and develop the effectiveness of four functional ecosystem parameters (bacterial uptake, phytoplankton photosynthesis, zooplankton grazing, and reproduction) as indicators of ecosystem health, (3) to determine the effects of complex effluents on controls of differing trophic state, (4) to examine the interactions between nutrient stimulation and the effects of toxicants in complex effluents, and (5) to relate inhibition and stimulation to the concentrations of specific metals and organics measured in the effluents. The results on the phytoplankton are presented in this paper.

Study Area

The River Raisin drains 2770 km² through predominantly agricultural southeastern Michigan and northwestern Ohio. Three cities and numerous small towns are located on the river with their associated industries and sewage treatment plants. The industries in the basin are generally suppliers for the automobile manufacturers in Detroit [15].

This study centered on the heavily industrialized Monroe Harbor area (Fig. 1). A Ford Motor Co. plant is located on the north bank where the river drains into Lake Erie. Mason Run is upstream from the plant and drains a wetland area heavily impacted with polychlorinated biphenyls (PCBs) [16]. On the opposite bank from the Ford plant is a Detroit Edison coal-fired power plant. Often a large portion of the volume of the river is drawn through the power plant

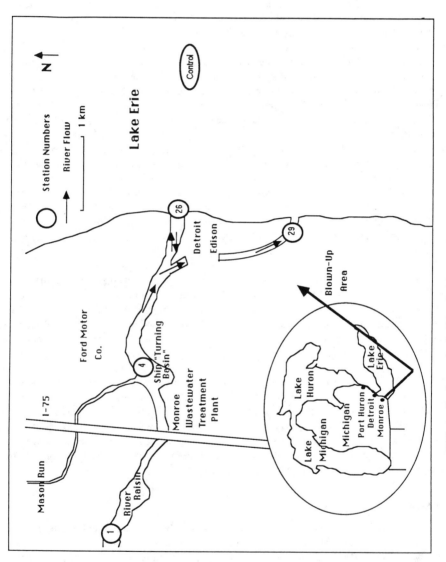

FIG. 1—*Study area in the River Raisin, Monroe Harbor, and Lake Erie.*

for cooling purposes and exits into Lake Erie south of the normal river mouth. A consequence of this divergence of the river is that Lake Erie water actually flows into the normal river mouth during peak periods of power production. Slightly upstream from Mason Run is the "turning basin," a widened area in the river where large Great Lakes bulk carriers turn, with much upheaval of sediments, to dock at the Detroit Edison plant. Upstream from the turning basin is the Monroe Wastewater Treatment Plant (WWTP), a secondary treatment facility.

During the ice-free portion of 1983 and 1984, a large number of stations were intensively surveyed for one week at approximately monthly intervals for chemical and physical variables and potential biological effects on the Lake Erie ecosystem. The stations included inflows and outflows to the River Raisin from the WWTP, the Ford Motor Co. plant, and the Detroit Edison plant and also sites placed throughout the lower river area in Monroe Harbor, in the river plume in the lake, and in the lake outside the mixing zone of the river [17]. The plume extends southward from the river mouth and the Detroit Edison outlet, its exact position dependent on local climatological conditions. In the summer of 1984 sampling every two weeks was concentrated on four stations (Fig. 1). Station 1 was approximately 1.6 km upstream from the WWTP, and thus was not impacted by the heavy industry in the harbor area but received background levels of pollutants from upstream industry, sewage treatment plants, and nonpoint agricultural sources. Station 4 was in the turning basin, downstream from the WWTP and Mason Run. Station 26 was at the original river mouth and could be comprised, depending on the operations of the Detroit Edison power plant, of either river water or Lake Erie water. Station 29 was at the mouth of the outlet of the Detroit Edison plant and consisted of river water that had passed through the power plant. This paper concentrates on the results from these four stations studied intensively over two years.

Experimental Procedure

A complex effluent test was used to determine the effects of the River Raisin station waters (that is, effluents) on the Lake Erie ecosystem. In 1983, water from various river stations was mixed in plastic bottles with Lake Erie water at concentrations of 10, 25, and 50% effluent and brought to a final volume of 200 mL. The rate of photosynthesis in the mixed samples was compared with that of a control sample of Lake Erie water only. The Lake Erie water was collected approximately 1.6 km east of the mouth, well outside the river plume.

In 1984, the effects of the River Raisin on a more oligotrophic lake were determined using Lake Huron water as a control. The untreated Lake Huron water was obtained from the intake pipe for the city of Detroit, north of Port Huron (Fig. 1). A 100% effluent sample was added for all stations.

Photosynthesis was measured as carbon uptake by the carbon-14 (^{14}C)–bicarbonate method [18]. Each 200-mL sample was injected with 5 μCi and incubated *in situ* at 1-m depth for approximately 4 h. Two 100-mL subsamples from each bottle were subsequently filtered onto 0.45-μm Gelman filters. The filters were each put into 5-mL Beckman high-performance (HpB) scintillation cocktail and were counted on a Beckman 1800 liquid scintillation counter. The external standard ratio method was used to arrive at the disintegrations per minute (DPMs). The average of the 2 100-mL subsamples was used for all analyses.

The U.S. Environmental Protection Agency (EPA) Large Lakes Laboratory at Grosse Ile, Michigan, and the Cranbrook Institute of Science at Bloomfield, Michigan, performed extensive chemical and physical analyses of the station effluents and controls. Statistical analyses were conducted by the U.S. EPA Large Lakes Laboratory and D. M. DiToro and his colleagues at Manhattan College.

Results

In 1983, photosynthesis in the Lake Erie control averaged 97.8 mg carbon/m^3/h [standard error (SE) = 15.8; n = 13] with a range from 30.6 to 228.2. In 1984, photosynthesis values in the Lake Huron control averaged 17.9 mg carbon/m^3/h (SE = 4.6; n = 6) with a range from 3.9 to 33.7. The photosynthesis rate of the Lake Huron control was significantly lower than that of the Lake Erie control (t' = 4.84; P < 0.001), using a Student's t-test for unequal population variances [19]. This indicates the more oligotrophic nature of the Lake Huron control. The photosynthesis rates in complex effluent samples in 1983 ranged from 3.6 to 510.5 mg carbon/m^3/h; in mixed effluent samples in 1984 the rates ranged from 4.5 to 591.2 mg carbon/m^3/h; and in the 100% station water in 1984 they ranged from 15.0 to 2099.6 mg carbon/m^3/h. The wide ranges indicate the variety of areas in which the stations were located, as well as normal seasonal variation in photosynthesis. All further values are reported as stimulation or inhibition of the complex effluent sample relative to the control value (100% Lake Erie water in 1983 or Lake Huron water in 1984) or to an expected value (to be defined further on).

In 1983, with Lake Erie water as the control, no clear trend was evident in inhibition or stimulation of photosynthesis with increasing percentages of Station 1 water over the various sampling dates (Fig. 2a). In 1984, with Lake Huron water substituted as the control, massive stimulation occurred with increasing percentages of Station 1 water (Fig. 2b). The river water contained not only increased amounts of nutrients and toxicants in relation to the Lake Huron control, but also a much larger biomass of phytoplankton, as indicated by the very high photosynthesis values of the 100% Station 1 water in relation to the control. Lake Erie is highly eutrophic itself, and thus the overriding effect of biomass in the complex effluent tests was not evident in 1983. No 100% effluent samples were run in 1983, so that the difference in algal biomass between the Lake Erie control and the station waters could not be directly compared.

In order to adjust for the differences in biomass between the complex effluents and the Lake Huron control, a weighted value for carbon-14 (^{14}C) uptake was obtained with the following equation

$$[\text{Carbon uptake of control water (mg C/m}^3/\text{h)} \times \% \text{ control water}]$$
$$+ [\text{carbon uptake of station water (mg C/m}^3/\text{h)} \times \% \text{ station water}]$$
$$= \text{expected carbon uptake with no inhibition or stimulation}$$

This weighted number was used as an expected value for no inhibition or stimulation occurring in the complex effluent test, instead of using Lake Huron water with its low biomass of phytoplankton and, thus, low photosynthesis rate. This manipulation of the data is actually nothing more than looking for discrepancies in an expected linear trend or slope in photosynthesis with increasing dilution of control water with station water.

When the Station 1 values were corrected for algal biomass, the average photosynthetic response over all dates in 1984 showed slight stimulation for samples that contained between 10 and 25% effluent, with no difference between the averages for the 25 and 50% dilutions (Fig. 2c). The responses on the individual sampling dates had considerable variation around this average trend.

In 1983 with eutrophic Lake Erie as the control, samples from Station 26, located at the river mouth, showed, on average, stimulation at concentrations containing between 10 and 25% effluent, and slight inhibition at concentrations of between 25 and 50% effluent (Fig. 3a). Again, much variation existed around the average trend. With the oligotrophic Lake Huron control in 1984, massive stimulation was observed with increasing percentages of effluent when the data were uncorrected for algal biomass (Fig. 3b). The stimulation was especially evident at sampling dates with high ambient water temperatures. When the 1984 data at Station 26 were corrected for algal biomass, determined by carbon-14 uptake, a very different result emerged (Fig. 3c). Inhibition occurred at all stations and dilutions, except for the 10% effluent on 30 May 1984 and the 25 and 50% effluent on 11 May. Very little effect of increasing the percentage of effluents in the samples was evident beyond the initial inhibition caused by the 10% effluent.

A similar response to that found at Station 26 occurred at Station 4, in the turning basin, and Station 29, the outflow of the Detroit Edison power plant. No clear trend in inhibition or stimu-

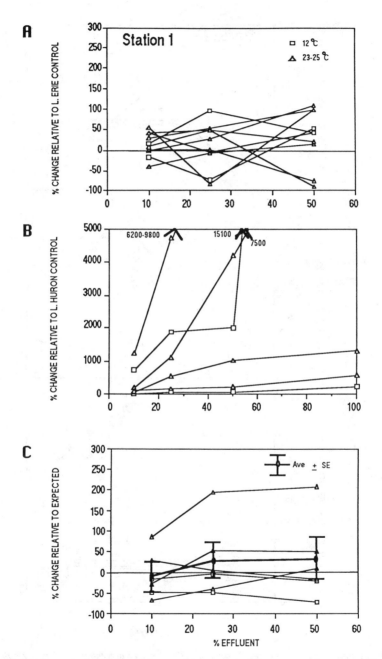

FIG. 2—*Inhibition and stimulation of photosynthesis at various percentages of complex effluents from Station 1 relative to controls:* (a) *Lake Erie control (1983),* (b) *Lake Huron control (1984), and* (c) *an expected value (corrected for algal biomass) using Lake Huron as the control (1984).*

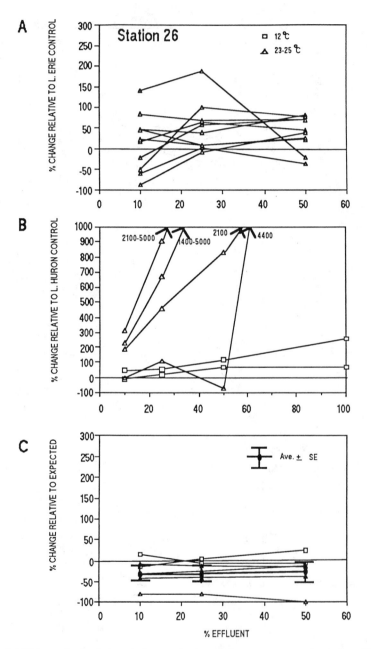

FIG. 3—*Inhibition and stimulation of photosynthesis at various percentages of complex effluents from Station 26 relative to controls:* (a) *Lake Erie control (1983),* (b) *Lake Huron control (1984),* and (c) *an expected value (corrected for algal biomass) using Lake Huron as the control (1984).*

lation of photosynthesis was evident in 1983 when eutrophic Lake Erie water was used as the control. The uncorrected 1984 data, with the Lake Huron control used as a diluant, indicated massive stimulation of photosynthesis with increasing percentages of effluent. After adjusting the data for the relative biomass of algae as ^{14}C-uptake potential, inhibition was observed at all percentages of effluent (Fig. 4a). No dose-response relationship with increasing effluent percentage was evident.

A consistent response was found at the four intensively surveyed stations when the photosynthesis rates at each station were averaged over all dates in 1984 (Fig. 4a). The uncorrected data all indicate massive stimulation with increasing effluent percentage. Station 1, however, located upstream from the major industry in Monroe Harbor, was more stimulated than all the downstream stations. On a comparative basis, this would indicate some inhibitory effect of the inputs into the river from the harbor industries. The spatial discrepancy between Station 1 and the harbor stations is even more evident when the values were corrected for algal biomass. Station 1 was stimulated in relation to the Lake Huron control water at the 25 and 50% effluent concen-

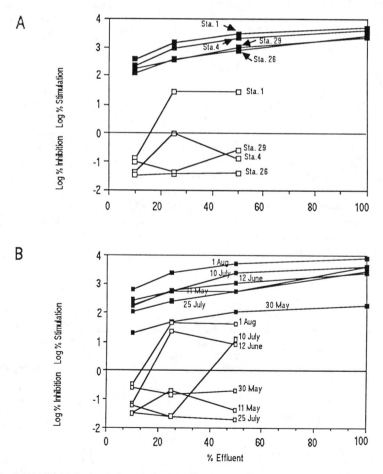

FIG. 4—*Inhibition and stimulation of photosynthesis in relation to both a Lake Huron control (solid squares) and an expected value (open squares):* (a) *averaged over all sampling dates in 1984 for four intensively studied stations and* (b) *averaged for each sampling date over the four stations.*

trations, while all other stations were inhibited at all effluent levels, with the exception of no change at the 25% effluent concentration for Station 4.

When the averages over all stations for each sampling date in 1984 are examined, the uncorrected data again show massive stimulation of photosynthesis with increasing percentages of station water on all dates except for 30 May 1984 (Fig. 4b). A more varied response occurred when the data were weighted for algal biomass, with both stimulation and inhibition occurring at different dates and dilutions (Fig. 4b).

Two analyses of covariance [19] were done on the 1984 data, with the photosynthesis rate (that is, the response variable) corrected and uncorrected for algal biomass, dilution percentage as the independent variable, and chemical and physical variables as the covariates (Table 1). Without the correction for algal biomass, the effects on photosynthesis rate of dissolved oxygen, conductivity, alkalinity, and total suspended solids were highly significant. The single toxicant to show an effect was dissolved chromium, which was almost statistically significant, although all the other toxicants (for example, zinc, copper, and chlorine) present in the station effluents were also tested for significance. The overall F value of the analysis of covariance was highly significant, and these factors explained almost half the variance observed in the photosynthetic response. When the correction for algal biomass was performed, the effect of total suspended solids remained highly significant, while the effect of dissolved oxygen was almost significant. No toxicant showed a significant effect on photosynthesis. The overall F value in the analysis of covariance remained highly significant, but now only 35% of the variance in photosynthesis was explained.

Discussion

Potential contributions of complex effluents to the photosynthetic response in a receiving body of water include the biomass and species composition of algae added, nutrient supplements, and toxicants. As this study was concerned with the integrated responses of Lake Erie and Lake Huron to complex effluents, we did not partition the potential interactive components of the River Raisin effluent. Where multiple stresses act simultaneously upon an ecosystem, partitioning of the effects may give misleading results. For example, phytoplankton may react differently to toxicants, depending on the nutrient conditions under which they are tested. Experimental procedures necessary to separate individual effects may also produce an artificial environment. For example, filtration of waters to remove phytoplankton will also remove particulates; yet, many metal and organic toxicants have been shown to bind strongly to particulates. It is necessary to maintain as realistic experimental conditions as possible where the effects of

TABLE 1—*Comparison of primary production changes with and without 100% correction using analysis of covariance.*

Variable	Without Correction		With Correction	
	F Value	Significance, P	F Value	Significance, P
Dilution	9.05	0.0001	12.87	0.0001
Dissolved oxygen	25.41	0.0001	3.54	0.0629
Conductivity	10.03	0.0020	2.94	0.0898
Alkalinity	6.22	0.0143	0.30	0.5854
Total suspended solids	18.36	0.0001	6.39	0.0131
Dissolved chromium	3.74	0.0559	1.48	0.2260
Overall	11.36	0.0001	6.66	0.0001
	$R^2 = 0.48$		$R^2 = 0.35$	

complex effluents on a receiving body of water are being studied, even though this may not allow ready separation of confounding components within the effluent.

There are potential problems when using ^{14}C-uptake as both an estimate of biomass and the response variable. An independent measurement of biomass would probably be preferred in future studies. Other measurements of biomass have their own limitations, however. Chlorophyll *a* content has been commonly used, but it has been shown to be strongly dependent on the nutrient state of the algae [20–22]. When using complex effluent tests with the control and effluent having very different trophic states, significant errors can result. Probably the most unbiased measurement of biomass is the ash-free dry weight [23], but much sample manipulation and time are required in preparation. On large-scale surveys involving many stations, such as this study, the logistics and time constraints of sampling become major obstacles. Also in systems with much detritus, such as rivers, much of the organic matter measured as ash-free dry weight may not be actively photosynthesizing algae. A review of the ^{14}C-uptake method of measuring photosynthesis, including the controversy over whether it estimates net or gross primary productivity, has been done by Peterson [24]. Problems with the ^{14}C-uptake technique in nutrient-enriched algal cultures have been indicated by Lean and Pick [25].

Organic contaminants in the River Raisin have been measured at significant concentrations [16], but the use of plastic bottles for incubations makes the actual sample concentrations of questionable accuracy because of the hydrophobic nature of the organics. Many bottles were simultaneously incubated *in situ,* and plastic was necessary to prevent breakage due to turbulence and subsequent release of low-level isotopes.

The 1984 corrected data indicate that the River Raisin would potentially inhibit photosynthesis in an oligotrophic receiving body of water. This is beyond any stimulatory effect of the river water from nutrient additions. A downstream gradient in toxicity existed, with the upstream station stimulatory at higher concentrations of effluent while the harbor stations were inhibitory. No similar conclusions can be drawn from the 1983 data when eutrophic Lake Erie water was used as a control, but no 100% correction was run and, thus, no correction for algal biomass could be done. In 1983, life tables were run on *Ceriodaphnia* and feeding trials on zooplankton isolated from Lake Erie. These indicated that the river stations were toxic to zooplankton in mixed effluent tests when Lake Erie water was used as a control [17,26,27].

It is interesting that, when an analysis of covariance was performed on the 1984 corrected photosynthesis data, less overall variance was explained than with the uncorrected data. Nevertheless, the corrected data may be more representative of the actual effect of the mixed effluents from the River Raisin, at least on the phytoplankton of an oligotrophic receiving body of water if not on Lake Erie [28]. No single toxicant had a statistically significant effect on photosynthesis, yet the corrected data indicate inhibition in the harbor stations. This discrepancy could have occurred for a number of reasons. It is possible that the pollutant causing the inhibition was simply not measured, although the U.S. EPA Large Lakes Research Station coordinated a large-scale environmental chemistry study coincident with the biological testing of the river's impact. In addition, many toxicants were present in the river, both heavy metals and organic compounds, and it could be that these had complex interactions so that no single chemical caused a significant effect on photosynthesis. It does appear that rivers with heavy loads of nutrients and toxicants can have significant effects on such basic trophic level responses as photosynthesis in the Great Lakes. As primary production is the base of the trophic energy pyramid, disruption of photosynthetic activity could potentially have profound ecological consequences.

Large-scale, multidisciplinary projects such as this one are extremely expensive, time-consuming ventures, and obviously in this study many questions remain unanswered concerning our ability to measure ecosystem-wide impacts of toxicants. It is important, though, to determine whether trophic level responses can be better correlated with specific toxicants in further studies. Other aquatic systems with a myriad of chemical pollutants, all at subthreshold levels, may provide a similar lack of correlation with specific environmental perturbations. Nevertheless, if ecosystems do indeed have unique system-level properties and if the ultimate goal of environmental regulation is to protect the entire ecosystem, research must be focused on actual ecosystems with their inherent complexities using a holistic research framework.

Conclusions

This paper has examined the interactions between nutrients and toxicants on photosynthesis in a complex effluent test for the Monroe Harbor area of the River Raisin. A varied response occurred when eutrophic Lake Erie water was used as the control. When Lake Huron water was substituted as the control, massive stimulation of photosynthesis apparently resulted. Yet, after weighting the photosynthesis values for the algal biomass, a downstream gradient in toxicity was evident. The upstream station caused stimulation of complex effluent samples, while the harbor stations caused inhibition. An analysis of covariance indicated that no single pollutant caused the inhibition found in the Monroe Harbor area.

Results suggest that, when an oligotrophic control is used with eutrophic effluents, a weighting factor for the relative biomass of algae must be used. Predictable trends occurred when this was done with the Lake Huron control. Further site-specific, ecosystem-level testing of environmental perturbations is called for.

Acknowledgments

The authors would like to thank the personnel of the U.S. EPA Large Lakes Laboratory in Grosse Ile, Michigan, and the Cranbrook Institute, in Bloomfield, Michigan, for collection of the samples and chemical and statistical analyses. D. M. DiToro and his colleagues at Manhattan College provided further analysis of the results. C. J. Richardson and R. T. Di Giulio provided valuable criticism of this manuscript. This work was supported by a grant from the U.S. EPA (CR810775) to D. C. McNaught.

References

[1] Loewengart, G. and Maki, A. W. in *Multispecies Toxicity Testing,* J. Cairns, Jr., Ed., Pergamon Press, New York, 1985, Chapter 1, pp. 1–12.
[2] Mount, D. I. in *Multispecies Toxicity Testing,* J. Cairns, Jr., Ed., Pergamon Press, New York, 1985, Chapter 2, pp. 13–18.
[3] Tebo, L. B., Jr., in *Multispecies Toxicity Testing,* J. Cairns, Jr., Ed., Pergamon Press, New York, 1985, Chapter 3, pp. 19–26.
[4] Giesy, J. P. in *Aquatic Toxicology and Hazard Assessment: Eighth Symposium, ASTM STP 891,* R. C. Bahner and D. J. Hansen, Eds., American Society for Testing and Materials, Philadelphia, 1986, pp. 67–77.
[5] Cairns, J., Jr., *Marine Environmental Research,* Vol. 3, 1980, pp. 157–159.
[6] Cairns, J., Jr., *Water Research,* Vol. 15, 1981, pp. 941–952.
[7] Kimball, K. D. and Levin, S. A., *BioScience,* Vol. 35, 1985, pp. 165–171.
[8] Cairns, J., Jr., *Hydrobiologia,* Vol. 100, 1983, pp. 47–57.
[9] Matthews, R. A., Buikema, A. L., Cairns, J. Jr., and Rodgers, J. H., *Water Research,* Vol. 16, 1982, pp. 129–139.
[10] Bormann, F. H. and Likens, G. E., *Pattern and Process in a Forested Ecosystem,* Springer-Verlag, New York, 1979.
[11] Odum, E. P., *Science,* Vol. 164, 1969, pp. 262–270.
[12] Odum, E. P., Finn, J. T., and Franz, E. H., *BioScience,* Vol. 29, 1979, pp. 349–352.
[13] Odum, E. P., *BioScience,* Vol. 35, 1985, pp. 419–422.
[14] International Joint Commission on the Great Lakes, Annual Report, Windsor, Canada, 1985.
[15] "River Quality in the Raisin River Basin," Michigan Department of Natural Resources, Lansing, MI, 1979.
[16] Filkins, J. C., Mullin, M. D., and Smith, V. E., "The Surficial and Vertical Distribution of Polychlorinated Biphenyls in the Sediment of the Lower River Raisin," *Proceedings,* 28th Conference, International Association of Great Lakes Research, Milwaukee, WI, 1985.

[17] McNaught, D. C., Bridgham, S. D., and Meadows, C., "Effects of Contaminants in the River Raisin on Ecosystem Function of the Bacteria, Phytoplankton, and Zooplankton," U.S. Environmental Protection Agency, Duluth, MN, 1985.

[18] Vollenweider, R. A. in *Primary Production in Aquatic Environments*, Blackwell Scientific, Oxford, England, 1969, pp. 41-125.

[19] Steel, R. G. D. and Torrie, J. H., *Principles and Procedures of Statistics: A Biometrical Approach*, 2nd ed, McGraw-Hill, New York, 1980.

[20] Droop, M. R. in *Algal Physiology and Biochemistry, Botanical Monographs*, Vol. 10, W. D. P. Stewart, Ed., University of California Press, Berkeley, CA, 1974, Chapter 19, pp. 530-559.

[21] Fogg, G. E., *Symposia of the Society for Experimental Biology*, Vol. 13, 1959, pp. 106-125.

[22] Skoglund, L. and Jensen, A., *Journal of Experimental Marine Biology and Ecology*, Vol. 21, 1976, pp. 169-178.

[23] Cole, G. A., *Textbook of Limnology*, 2nd ed., C. V. Mosby, St. Louis, 1979, pp. 92-94.

[24] Peterson, B. J., *Annual Review of Ecology and Systematics*, Vol. 11, 1980, pp. 359-385.

[25] Lean, D. R. S. and Pick, F. R., *Limnology and Oceanography*, Vol. 26, 1981, pp. 1001-1019.

[26] McNaught, D. C., Bridgham, S. D., and Meadows, C., "Effects of Complex Effluents from the River Raisin on Zooplankton Grazing in Lake Erie," this publication, pp. 128-137.

[27] McNaught, D. C. and Mount, D. I., *Aquatic Toxicology and Hazard Assessment: Eighth Symposium, ASTM STP 891*, R. C. Bahner and D. J. Hansen, Eds., American Society for Testing and Materials, Philadelphia, 1986, pp. 375-381.

[28] Bridgham, S. D., McNaught, D. C., Meadows, C., Rygwelski, K., Dolan, D., and Gessner, M., "Factors Responsible for the Inhibition or Stimulation of Two Great Lakes Ecosystems," *Proceedings*, 28th Conference on Great Lakes Research, International Association of Great Lakes Research, Milwaukee, WI, 1985.

J. David Yount[1] *and Lyle J. Shannon*[2]

State Changes in Laboratory Microecosystems in Response to Chemicals from Three Structural Groups

REFERENCE: Yount, J. D. and Shannon, L. J., "State Changes in Laboratory Microecosystems in Response to Chemicals from Three Structural Groups," *Functional Testing of Aquatic Biota for Estimating Hazards of Chemicals, ASTM STP 988,* J. Cairns, Jr., and J. R. Pratt, Eds., American Society for Testing and Materials, Philadelphia, 1988, pp. 86–96.

ABSTRACT: Generic mixed-flask culture microecosystems derived from small lake and pond planktonic communities were used to evaluate the effects of selected alcohols, aniline derivatives, and aromatic amides on pH and dissolved oxygen—ecosystem state variables which reflect energy flow and nutrient cycling processes within the systems.

Using changes in these variables as indicators of effect, the relative toxicity rankings of compounds within each group were determined and compared with the toxicity rankings established by fathead minnow acute toxicity tests. The ecosystem-level relative toxicity of three alcohols agreed with the toxicity rankings based on fathead minnow 96-h lowest-observed-effect levels (LOELs). The toxicity rankings of aniline and three of its derivatives were similar to those for the alcohols. For both alcohols and anilines, the ecosystem state variables were more sensitive than the fathead minnow lethality for the least toxic members of the group. A group of aromatic amides, which included an inhibitor of photosynthesis and an uncoupler of oxidative phosphorylation, showed the most departure from the fathead minnow toxicity ranking. The amide compound that was least toxic to fathead minnows was most toxic at the ecosystem level.

These examples can be considered representative of situations that might be encountered in an early stage of ecosystem-level testing. For less well known toxicants, microcosm functional testing could assist in identifying chemicals that require more elaborate test procedures.

KEY WORDS: hazard evaluation, microcosms, laboratory microecosystems, toxicity testing, hazard ranking, ecosystem effects, alcohols, anilines, aromatic amides

This work was inspired, in part, by a critical review [1] in which aquatic and terrestrial laboratory multispecies test systems were evaluated for their utility in ecological hazard and risk assessment processes. Since then, numerous authors have addressed the need for methods for environmental effects assessment of chemicals at the community and ecosystem level [2-6]. Cairns [7] has also noted the lack of a systematically generated body of data to determine the usefulness of data generated at one level of biological organization in predicting responses at a higher level of organization. The authors of this paper believe that the work reported here contributes to such a body of data, and that our results provide insight into how ecosystem-level testing might be incorporated into a tier-testing scheme at an early stage.

Leffler [8] has developed a protocol to quantify the effects of chemicals on a few easily measured integrative properties of model ecosystems and to rank chemicals in order of the concentrations required to produce an observable effect. He has hypothesized that these rankings will be consistent among mixed cultures with differing species composition, even though the abso-

[1]Life scientist, U.S. Environmental Protection Agency, Environmental Research Laboratory, Duluth, MN 55804.
[2]Research fellow, Department of Biology, University of Minnesota, Duluth, MN 55812.

lute values of the measured variables may not be accurate predictors of effect levels in other mixed cultures or in natural ecosystems. Giddings [9] has observed that

the major criterion for such a laboratory system is its ability to generate rankings that are consistent with the actual hazard potential of the chemicals in nature, rather than its ability to simulate specific ecosystem effects. Test chemicals could be compared with selected standard reference chemicals to identify those with the greatest potential for environmental effects.

In 1983 we began a program to evaluate the utility of Leffler's microcosm testing protocol for routinely testing large numbers of chemicals. We were interested not only in the relative toxicity or hazard rankings produced by the test but also in its sensitivity in comparison with representative single-species tests. We used the extensive fathead minnow (*Pimephales promelas*) toxicity data base developed at the U.S. Environmental Protection Agency (EPA) Environmental Research Laboratory in Duluth (ERLD) to establish a rank order against which the relative toxicity to microcosm communities could be compared. This data base was generated to develop quantitative structure-activity relationships (QSAR) for toxicants with similar modes of action. From this data base we were therefore able to select three chemical groups containing compounds with different representative modes of action and widely differing toxicity. However, because our primary objective was to examine the relative toxicity rankings determined by ecosystem-level effects, we did not require that our test compounds have an extensive single-species toxicity data base. Therefore, several of the compounds tested had only fathead minnow toxicity data available.

Detailed analyses of these tests are summarized in other publications and reports [10–12], in which each chemical group is treated separately. The purpose of this paper is to take a broader look at the data and to consider differences between the relative hazard rankings for each chemical group. At this level of analysis the *experiment* becomes the full set of tests on a group of compounds and the *hazard ranking* itself becomes the *output* of an experiment, as outlined by Giddings [9]. These rankings are hypothesized by Giddings to be more consistent among different ecosystems, and hence more generalizable, than qualitative or quantitative predictions of effects would be. In this paper we address only the differences in relative hazard rankings generated by two laboratory tests: an ecosystem-level test and a single-species test. The hypothesis that hazard rankings will be consistent among different ecosystems remains to be tested.

Materials and Methods

Only ecosystem state variables were measured. Our justification for treating pH and dissolved oxygen (DO) as indicators of the state of the ecosystem (that is, as ecosystem state variables) is based on a paper by Schindler et al. [13]. Figure 1 presents a conceptual model of an ecosystem as an integrated biogeochemical system. In a system without biological activity, the geochemical matrix (the nutrient medium in our laboratory ecosystems) can be considered to be in a thermodynamic ground state, or equilibrium, which is completely characterized by its pH and oxidation-reduction (or redox) potential [14]. With the addition of an inoculum of biota comprising producers, consumers, and decomposers and a source of light energy, coupled anabolic and catabolic processes give rise to a nonequilibrium thermodynamic state determined by the ecological energy flow and nutrient cycling processes occurring within the system, which is now an eco-system. This biological activity, through electron and proton transfers, produces changes in the pH and redox potential of the aqueous phase, and the resulting normal range of these values can be considered to represent the unperturbed state of the system. Perturbations to the ecosystem, if they affect energy flow or nutrient cycling processes, will produce corresponding changes in pH and redox potential and move one or both out of the normal range. Recovery occurs when the pH and redox potential return to their normal range. Following Waide et al. [15], we as-

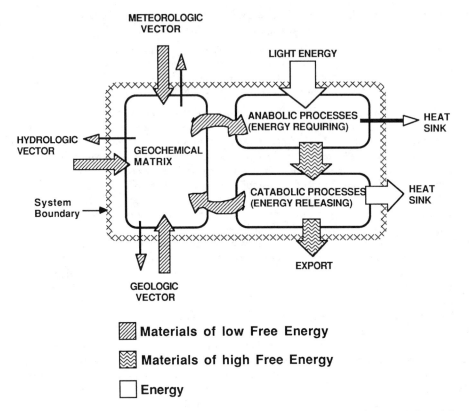

FIG. 1—*Conceptual model of an ecosystem as an integrated biogeochemical system (adapted from Ref 13).*

sumed that DO provided a suitable indicator of redox potential in aerobic systems. Therefore, we used pH and DO as integrators of the processes occurring within the microcosms and indicators of changes in those processes due to toxic chemical effects. In earlier work we found these measurements, which can be more sensitive than species counts [16], to be reliable indicators of effect.

Mixed-flask culture microcosms were developed according to the method of Leffler [8,17]. These microcosms consisted of 1-dm³ (litre) beakers containing 950 cm³ (mL) of Leffler's modification of Taub No. 36 medium [8,18] and 50 cm³ of a mixed culture inoculum. The inoculum was taken from a mature stock community consisting of zooplankton, algae, and other planktonic organisms collected from local lake and pond communities. Because we monitored only ecosystem-level variables, we were not concerned with species identification other than to verify that the inoculum contained the following minimum diversity within functional groups: two unicellular green algal species, one nitrogen-fixing blue-green algal species, one filamentous green algal species, one grazer species, one benthic detritivore species, bacteria, and protozoa. We set no upper limit on taxonomic diversity.

The microcosm containers were not covered, in order to allow free exchange of gases with the atmosphere. Evaporative losses provided sufficient free volume (at least 50 cm³) for weekly reinoculation to simulate natural immigration and to provide the potential for system recovery. Distilled deionized water was added when necessary to maintain a total volume of 1 dm³.

The microcosms were arranged in a randomized block design in a Percival I-60 environmental growth chamber at 293 K (20°C). Fluorescent light banks provided illumination on a 12-h light/12-h dark cycle. To guard against potential effects from nonuniform incubator conditions, all the blocks were rotated to a different shelf position twice weekly.

For each test, 40 microcosms were started and monitored for a six-week predose development phase. Prior to treatment, the 5 microcosms showing the greatest deviation from the group were culled from the set, so that the actual treatment design consisted of 35 microcosms divided into control and treatment groups each containing 5 replicates. These systems were then monitored for six weeks after treatment. Posttreatment monitoring was done on a schedule of decreasing frequency. Measurements were made on four days during the first week, two days during the second week, and once weekly thereafter.

Dissolved oxygen was determined electrochemically using an Orbisphere Model 2607 meter in early experiments and an Altex (Beckman) Model 0260 oxygen analyzer in later work. Measurements of pH were taken using a Fisher 525 meter attached to a junction box equipped with standard pH electrodes. Although no counts of population densities were made, qualitative observations of change in algal or zooplankton abundance were recorded.

The compounds tested were chosen from the extensive set of toxicity data on the fathead minnow *(Pimephales promelas)* developed for the QSAR program at the U.S. EPA Environmental Research Laboratory in Duluth, Minnesota.[3] Toxicity data on over 450 compounds in over 40 chemical categories provided a wide choice of toxicity levels and modes of action. Three chemical groups (Fig. 2) were chosen to represent distinctly different examples of possible groups of compounds with different environmental chemistry and modes of action.[4]

The treatment levels were chosen on the basis of the fathead minnow 96-h median lethal concentration (LC_{50}) values, microcosm range-finding tests, and solubility considerations. In selecting treatment levels, the reported fathead minnow LC_{50} was used as the median concentration. Extreme concentrations were set two orders of magnitude above and below this level, if solubility limits permitted, and two or three intermediate treatment levels were selected. These intermediate levels frequently differed by an order of magnitude. Acetone was used as a carrier solvent for most solid compounds and sparingly soluble liquids. Methanol was the solvent for 4-hexyloxyaniline, a solid, because it appeared to degrade in acetone. When a solvent was necessary, a volume not exceeding 10^{-4} volume/volume was added to all treated systems and to a set of solvent controls, thereby reducing the number of toxicant treatment levels from six to five. The microcosms were dosed once only, so that the toxicant levels presumably declined over time, and the systems had the potential for recovery.

The term "concentration" as used in this paper is not the aqueous phase concentration (except perhaps immediately after the toxicant is added) but the nominal concentration, that is, the total calculated amount of toxicant added per unit volume of the microcosm. When a toxicant was added to a microcosm test system, we were dosing the entire ecosystem, not specifically treating the organisms within the microcosm. In this respect microcosm tests are different from single-species or multiple-species exposure tests. Once a toxicant is added to a microcosm, it is quickly partitioned among the biota, the liquid phase, and other abiotic components of the system. Therefore, a measurement of toxicant in the liquid phase, unless taken immediately after addition, is of little value in determining the amount of toxicant in the system.

Dunnett's procedure [19] was used to determine which treatment groups were significantly different from the control. With this information, the lowest-observed-effect level (LOEL) was determined individually for each measured primary variable (morning DO, morning pH, evening DO, and evening pH) at each measurement time. The LOEL for the microcosm test as a whole was chosen as the lowest measured primary-variable LOEL during a specified time pe-

[3]Unpublished data base, EPA Environmental Research Laboratory at Duluth, Minnesota.
[4]The full names and Chemical Abstract Service (CAS) numbers for the chemicals tested are as follows: 1-decanol (CAS No. 112-30-1); 1-octanol (CAS No. 111-87-5); 1-hexanol (CAS No. 111-27-3); 2,3,5,6-tetrachloroaniline (CAS No. 3481-20-7); 4-hexyloxyaniline (CAS No. 39905-57-2); 2,6-diisopropylaniline (CAS No. 24544-04-5); aniline (CAS No. 62-53-3); 2-(octyloxy)acetanilide (CAS No. 55792-61-5); salicylanilide (CAS No. 87-17-2); diuron, or DCMU, or 3-(3,4-dichlorophenyl)-1,1-dimethyl urea (CAS No. 33-05-4).

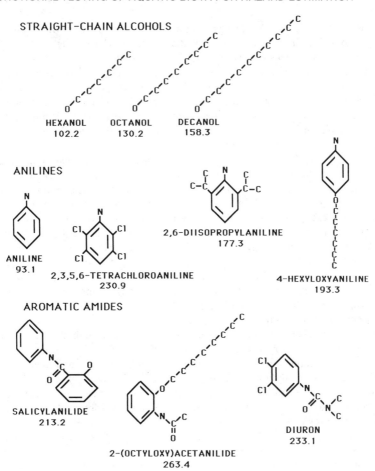

FIG. 2—*Skeletal structures and molecular weights of the test compounds.*

riod, either one to four days (96 h) or longer than four days. A combination of variables, such as diurnal oxygen gain or loss or the ratio of gain to loss, was generally less sensitive than the individual variables because of what we interpreted as the random addition of errors. In general, we have found that the use of the primary variables directly is, at least statistically, a more sensitive method.

A typical data set for a given variable is illustrated by Fig. 3, in which the difference between the mean of a measured variable and the control group mean is plotted versus time. Dunnett's least significant difference [*19*] is shown as a pair of lines above and below the zero deviation line which bound the shaded region of nonsignificant difference. This type of figure clearly shows the relationship between toxicant concentration and ecosystem dysfunction and recovery. Since we were making four separate statistical comparisons (morning and evening measurements for DO and pH) on each of ten measurements days (for a total of 40 tests), we would expect that two of these tests (5%) might show a significant difference just by chance at $p = 0.05$. To reduce this possibility, we required that statistically significant differences from controls occur on at least two *consecutive* sampling times (consecutive mornings or consecutive evenings) before we considered an effect to be real.

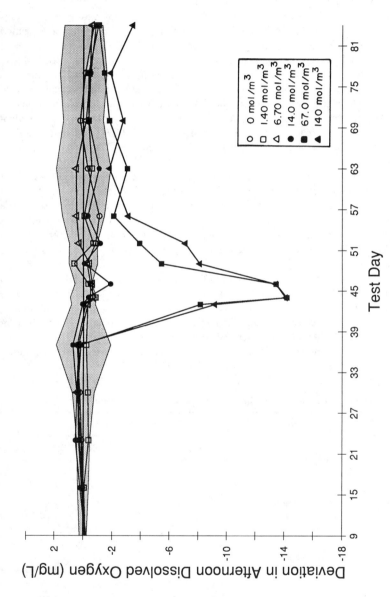

FIG. 3—*Typical data set, in which the deviation of a variable (afternoon DO) from the control is plotted versus the test day. Shaded area is the region of nonsignificant difference, from Dunnett's test (P ≤ 0.05) [19]. The toxicant, 4-hexyloxyaniline, added on Day 42, and measurements normally continued through Day 84. The DO deviations expressed in moles per cubic metre, are 31.25 × mg/L.*

Results

Since the compounds tested were selected from a structure-activity data base on the fathead minnow to represent different chemical groups and modes of action, the fathead minnow is the only species on which all of our test compounds have been evaluated. Accordingly, 96-h LC_{50} data from these tests were used to establish a rank order against which the relative toxicity to microcosm communities could be compared. The data from the three intragroup comparisons are summarized in Table 1. Within each group, the compounds are listed by their abbreviated names in the order of their relative toxicity to fathead minnows. The fathead minnow 96-h LOEL, rather than the LC_{50}, is listed to provide a more comparable indicator of the relative sensitivity of the microcosm test in comparison with the single-species test. The last two columns give the lowest-observed-effect ($P \leq 0.05$) levels for the microcosms determined from morning or evening DO or pH measurements. The LOELs are listed both for effects occurring during the first 96 h and for effects occurring between 4 and 42 days after treatment. The alcohols and tetrachloroaniline were tested twice, as previously reported [11,12]. Where the LOELs differed between tests, the lowest effect level is the one recorded here. The other compounds were tested once.

The relationships between the hazard rank order and relative sensitivity within these three chemical groups are summarized in Fig. 4. Here the 96-h LOELs listed in Table 1 are plotted in a dot chart on an identical log (base 10) scale for each group so that the relative toxicities of the chemicals in each group can be easily seen by noting the positions of the LOELs on the horizontal axis. The concentrations decrease from left to right so that the toxicity will increase toward the right of the graph. The microcosm 96-h LOELs are slightly offset above the corresponding fathead minnow LOELs.

In the alcohol group, the short-term ecosystem LOELs for decanol and octanol were close to the fathead minnow 96-h LOEL values in both absolute value and rank order. Hexanol, however, had a 96-h ecosystem LOEL of 97.9 moles/m³, considerably below the fathead minnow LOEL of 863 moles/m³. Therefore, although the hazard ranking is similar for the microcosm and fathead minnow tests, the microcosm test was more sensitive for hexanol, the least toxic member of the series. The long-term LOELs for the alcohols (Table 1) are equal to or higher

TABLE 1—Comparison of ecosystem versus fathead minnow hazard rankings for selected toxicants.[a]

Chemical	Fathead Minnow, 96-h LOEL, moles/m³	Microcosm	
		96-h LOEL, moles/m³	>96-h LOEL, moles/m³
Straight-chain alcohols			
Decanol	11	19	19
Octanol	68	77	77
Hexanol	863	97.9	979
Anilines			
Tetrachloroaniline	0.87	1.2	1.2
Hexyloxyaniline	14	6.7	14
Diisopropylaniline	59.8	8.5	8.5
Aniline	851	107	10.7
Aromatic amides			
Octyloxyacetanilide	<1.2	0.19	0.019
Salicylanilide	8.25	46.2	46.2
Diuron (DCMU)	24	0.12	0.012

[a]The chemicals are listed within a group in the order of fathead minnow toxicity. The numbers are the lowest-observed-effect level (LOEL) for both fathead minnows and microcosms. An effect is defined as lethality for fatheads or a statistically significant pH or DO change for microcosms. The full names of the chemicals are given in Footnote 4.

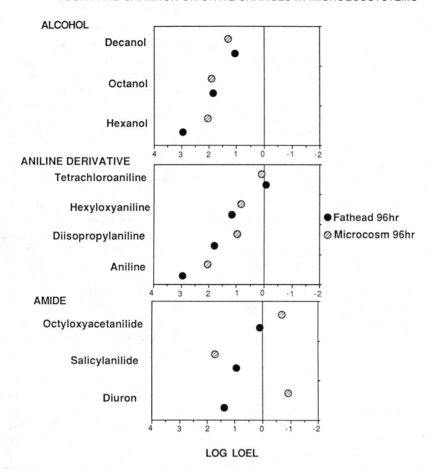

FIG. 4—*Comparison of the ecosystem and fathead minnow hazard rankings for selected toxicants. The toxicity increases toward the right of the graph. The LOEL values are expressed in moles per cubic metre. An effect is defined as lethality for fatheads or a pH or DO change for microcosms. The full names of the chemicals are given in Footnote 4.*[4]

than the short-term LOELs, since the groups with the lower treatment levels often recovered to control levels within four days of treatment.

With the exception of tetrachloroaniline, the aniline group of compounds was more toxic in 96 h to microcosms than to fathead minnows. The magnitude of this difference was especially large for diisopropylaniline and aniline. Based on long-term microcosm LOELs (Table 1) one could predict that there is not likely to be any significant difference in long-term toxicity between hexyloxyaniline, diisopropylaniline, and aniline in natural ecosystems.

The 96-h ecosystem responses to the substituted aromatic amides differed from the fathead minnow LOELs both in sensitivity and in rank order. During the first 96 h the microcosms were most sensitive to diuron, whereas the fathead minnows were least sensitive. Diuron caused short-term ecosystem effects at a level greater than two orders of magnitude below the fathead 96-h LOEL. Similarly, octyloxyacetanilide produced short-term ecosystem effects at one tenth of the fathead 96-h LOEL. The microcosms were less sensitive than the fathead minnow to salicylanilide during the first 96 h. With the exception of salicylanilide, long-term effects appeared in the microcosms in this chemical group at one tenth of the short-term effect levels.

Discussion

For general depressants or narcotics, which have similar effects on most or all organisms, we expected that ecosystem-level measurements of effect would rank the toxicities in the same order as single-species toxicity tests. Other compounds, belonging to groups with specific or multiple modes of action or having complex degradation pathways with potential toxic intermediates, would not necessarily be expected to exhibit a high correlation between ecosystem-level effects and any particular single-species toxicity test. This is because of the greater opportunity in ecosystems for complex interactions to occur between species in groups for which different modes of action are effective.

To paraphrase Giddings [9], it is the ranking of chemicals in the order of the concentrations required to produce an observable effect, not the observed effects themselves, that constitutes the output of this experimental design. Our model ecosystems were used to identify chemicals capable of disrupting ecosystem processes, not to specify which processes are disrupted or precisely how these effects might be manifested in natural systems. Since many single-species toxicity tests also have the objective of ranking chemicals by potential hazard, model ecosystems would be valuable primarily if they were more sensitive than conventional single-species tests or if they generated different rankings than those of conventional tests. If ecosystem-level tests merely echoed the results of simpler, more easily standardized tests, their use for screening chemicals would be questionable. Our results suggest that this relatively simple generic laboratory ecosystem test is promising on both counts.

Straight-chain alcohols represent general narcotics, acting to depress biological activity in all or most organisms. The ecosystem relative toxicities of three alcohols not only followed the single-species rankings but, with the exception of hexanol, were well within an order of magnitude of the fathead minnow 96-h LOELs.

Aniline and its derivatives not only are general narcotics but also exhibit other modes of action. In addition, toxic degradation intermediates may affect different taxonomic groups differentially [20]. This is in agreement with the ecosystem relative toxicities of the aniline group. Four-day LOELs from the microcosms ranked the aniline group in the same order as that obtained from the fathead minnow tests, but long-term effects in the microcosms produced a different ranking, possibly in response to toxic degradation products.

Many compounds in the aromatic amide group affect photosynthesis and thus would be expected to have a dramatic effect on ecosystems. Salicylanilide contains a phenolic group which causes it to act primarily as an uncoupler of oxidative phosphorylation [21]. Classification of these compounds by means of the amide functional group obviously has little relevance to the mode of action when the compounds contain other functional groups which are more important toxicologically. One would expect, therefore, that an ecosystem-level hazard ranking of chemicals in this group might be considerably different from a hazard ranking based on any particular single-species test. Diuron, the amide least toxic to fathead minnows, is a potent inhibitor of photosynthesis and was, in fact, the most toxic of all the compounds tested in microcosms. Octyloxyacetanilide, whose mode of action is not known to us at present, was nearly as toxic at the ecosystem level as diuron, whereas the fathead minnow toxicity of these two compounds differed by nearly two orders of magnitude.

These examples can be considered typical of situations that might be encountered in an early stage of ecosystem-level testing and illustrate how, for less well-known toxicants, microcosm functional testing could assist in identifying chemicals that would require more elaborate test procedures. In such cases, ecosystem-level testing of a new compound or several members of a class of compounds, perhaps in conjunction with an appropriate reference chemical or benchmark, can provide a trigger for initiating additional testing. We believe that these results support simultaneous toxicity testing at different levels of biological organization [7]. For example, if a chemical manufacturer is developing a new product in a chemical structure group about which little information exists for predicting its toxic mode of action, a relatively simple ecosystem-level test including one or more benchmark chemicals could indicate not only whether additional tests are desirable but also what kind of additional tests might be needed. Although it is not an essential part of this test protocol, qualitative or semi-quantitative observations of the early effects of toxicants on component trophic groups in the microcosms can also point toward additional tests on certain taxa.

Provided that measurements are restricted to easily measured characteristics of the state of the whole system, such as pH and DO, measuring ecosystem-level effects in the laboratory is not inherently more difficult than measuring single-species effects. The six-week development period need not be considered part of the test, and is analogous to the acclimation period for organisms brought in from the field or to the time required to produce the desired life stage for testing. At the most intense level of testing, we had three tests under way simultaneously: one in the predose phase, one in the two-week postdose period, in which the measurements were made most frequently, and one in the last four weeks of observation, during which measurements were made only once per week.

More relevant, however, is the effort required to obtain ecosystem-level hazard rankings of chemicals by other means. In this regard, laboratory ecosystems are definitely less labor-intensive than tests in outdoor ecosystems, which in many cases are not even feasible to perform.

Although we consider these results to provide support for ecosystem-level testing, we do not yet consider the results to be conclusive. There is a need for additional testing with different groups of compounds, for additional replication of tests on the same compounds, and for studies of the influence of aspects of the test protocol, such as media formulation and microcosm development time. Additional field and laboratory ecosystem tests are also needed to test the hypothesis that relative hazard rankings are consistent over different aquatic ecosystems.

Acknowledgments

The authors gratefully acknowledge the participation of Carl Mach, Terry Flum, Charles Walbridge, and Daniel Fitzsimmons in the collection of data and other essential laboratory activities. Dr. Michael Harrass worked with our group during the first year of the project. Dr. Jack Hargis, of the University of Minnesota at Duluth, participated in the initial conception of the research but did not live to see the fruits of his efforts. Laboratory space, supplies, and equipment were largely supplied by the U.S. Environmental Protection Agency, Environmental Research Laboratory in Duluth (Minnesota) and the support of non-EPA personnel was provided through EPA Cooperative Agreement No. CR810741.

References

[1] Hammons, A. S., Ed., *Methods for Ecological Toxicology: A Critical Review of Laboratory Multispecies Tests*, Ann Arbor Science, Ann Arbor, MI, 1981.
[2] Cairns, J., Jr., "Are Single Species Toxicity Tests Alone Adequate for Estimating Environmental Hazard?" *Hydrobiologia*, Vol. 100, 1983, pp. 47–57.
[3] Cairns, J., Jr., *Multispecies Toxicity Testing*, SETAC Special Publications Series, Pergamon Press, New York, 1985.
[4] Giddings, J. M., "Microcosms for Assessment of Chemical Effects on the Properties of Aquatic Ecosystems," *Hazard Assessment of Chemicals: Current Developments*, Vol. 2, J. Saxena, Ed., Academic Press, New York, 1983, pp. 45–94.
[5] Levin, S. A. and Kimball, K. D., "New Perspectives in Ecotoxicology," *Environmental Management*, Vol. 8, 1984, pp. 375–442.
[6] Kimball, K. D. and Levin, S. A., "Limitations of Laboratory Bioassays: The Need for Ecosystem-Level Testing," *Bioscience*, Vol. 35, 1985, pp. 165–171.
[7] Cairns, J., Jr., "The Case for Simultaneous Toxicity Testing at Different Levels of Biological Organization," *Aquatic Toxicology and Hazard Assessment: Sixth Symposium, ASTM STP 802*, W. E. Bishop, R. E. Cardwell, and B. B. Heidolph, Eds., American Society for Testing and Materials, Philadelphia, 1983, pp. 111–127.

[8] Leffler, J. W., "The Use of Self-Selected, Generic Aquatic Microcosms for Pollution Effects Assessment," *Concepts in Marine Pollution Measurements*, H. H. White, Ed., Maryland Sea Grant College, University of Maryland, College Park, MD, 1984, pp. 139–157.

[9] Giddings, J. M., "Laboratory Tests for Chemical Effects on Aquatic Population Interactions and Ecosystem Properties," *Methods for Ecological Toxicology: A Critical Review of Laboratory Multispecies Tests*, A. S. Hammons, Ed., Ann Arbor Science, Ann Arbor, MI, 1981, pp. 23–91.

[10] Flum, T. J. and Shannon, L. J., "The Effects of Three Related Amides on Microecosystem Stability," *Ecotoxicology and Environmental Safety*, Vol. 13, 1987, pp. 239–252.

[11] Yount, J. D. and Shannon, L. J., "Effects of Aniline and Three Derivatives on Laboratory Microecosystems," *Environmental Toxicology and Chemistry*, Vol. 6, 1987, pp. 463–468.

[12] Harrass, M. C., Shannon, L. J., and Yount, J. D., "Effects of Straight-Chain Alcohols on Microcosm Communities and Comparison with Fathead Minnow Toxicities," internal report, U.S. EPA Environmental Research Laboratory, Duluth, MN, 1987.

[13] Schindler, J. E., Waide, J. B., Waldron, M. C., Hains, J. J., Schreiner, S. P., Freedman, M. L., Benz, S. L., Pettigrew, D. R., Schissel, L. A., and Clark, P. J., "A Microcosm Approach to the Study of Biogeochemical Systems: 1. Theoretical Rationale," *Microcosms in Ecological Research*, J. P. Giesy, Ed., DOE Symposium Series 52, CONF-781101, U.S. Department of Energy, Washington, DC, 1980, pp. 192–203.

[14] Stumm, W. and Morgan, J. J., *Aquatic Chemistry: Introduction Emphasizing Chemical Equilibria in Natural Waters*, Wiley, New York, 1970.

[15] Waide, J. B., Schindler, J. E., Waldron, M. C., Hains, J. J., Schreiner, S. P., Freedman, M. L., Benz, S. L., Pettigrew, D. R., Schissel, L. A., and Clark, P. J., "A Microcosm Approach to the Study of Biogeochemical Systems: 2. Responses of Aquatic Laboratory Microcosms to Physical, Chemical, and Biological Perturbations," *Microcosms in Ecological Research*, J. P. Giesy, Ed., DOE Symposium Series 52, CONF-781101, U.S. Department of Energy, Washington, DC, 1980, pp. 204–223.

[16] Shannon, L. J., Harrass, M. C., Yount, J. D., and Walbridge, C. T., "A Comparison of Mixed Flask Culture and Standardized Laboratory Model Ecosystems for Toxicity Testing," *Community Toxicity Testing, ASTM STP 920*, John Cairns, Jr., Ed., American Society for Testing and Materials, Philadelphia, 1986, pp. 135–157.

[17] Leffler, J. W., "Tentative Protocol of an Aquatic Microcosm Screening Test for Evaluating Ecosystem-Level Effects of Chemicals," developed under Battelle Columbus Laboratories Subcontract T6411 (7197), Ferrum College, Ferrum, VA, 1981.

[18] Taub, F. B. and Dollar, A. M., "A *Chlorella-Daphnia* Food Chain Study: The Design of a Compatible, Chemically Defined Culture Medium," *Limnology and Oceanography*, Vol. 9, 1964, pp. 61–74.

[19] Steel, R. G. D. and Torrie, J. H., *Principles and Procedures of Statistics*, McGraw-Hill, New York, 1960.

[20] Aoki, K., Shinke, R., and Nishira, H., "Metabolism of Aniline by *Rhodococcus erythropolis* AN-13," *Agricultural and Biological Chemistry*, Vol. 47, 1983, pp. 1611–1616.

[21] Williamson, R. L. and Metcalf, R. L., "Salicylanilides: A New Group of Active Uncouplers of Oxidative Phosphorylation," *Science*, Vol. 158, 1967, pp. 1694–1695.

Cecilia Lindblad,[1] *Ulrik Kautsky,*[1] *and Nils Kautsky*[1]

An *In Situ* System for Evaluating Effects of Toxicants on the Metabolism of Littoral Communities

REFERENCE: Lindblad, C., Kautsky, U., and Kautsky, N., **"An *In Situ* System for Evaluating Effects of Toxicants on the Metabolism of Littoral Communities,"** *Functional Testing of Aquatic Biota for Estimating Hazards of Chemicals, ASTM STP 988,* J. Cairns, Jr., and J. R. Pratt, Eds., American Society for Testing and Materials, Philadelphia, 1988, pp. 97–105.

ABSTRACT: A portable *in situ* continuous-flow apparatus for measuring metabolic activities of littoral ecosystems is described. Oxygen, temperature, light, and flow rate are continuously monitored by a microcomputer that redirects the water flow through the measurement chamber from acrylic jars containing the communities. The apparatus is especially designed for experimental purposes, since it is possible to alter the incoming waters by adding various toxicants. The application of the method is presented by an example showing the response of a filter-feeding community to cadmium.

The community showed statistically significant changes in respiration (oxygen) and excretion (nitrogen or phosphorus) as a result of treatment. Calculated ratios—such as the oxygen/nitrogen ratio, oxygen/phosphorus ratio, and perturbation index (PI), which is the relative change due to treatment, taking into account changes in the controls—were more sensitive measurements of sublethal stress than specific values of oxygen, nitrogen, or phosphorus. These indexes are independent of biomass, and PI normalizes synchronic changes in the environment due to light, temperature, and salinity variations, which affect both treated and control systems equally. The design of the system allows the communities to be placed back in the environment after being tested and also allows repeated measurements to be made of possible delayed effects or recovery from perturbations. In this paper, the authors introduce a new measurement of disturbance, the absolute disturbance index (ADI). ADI is the absolute distance between the metabolic activity of an undisturbed system and a disturbed point in a multidimensional space where each dimension represents the PI of a measured parameter.

KEY WORDS: aquatic ecosystems, functional testing, hazard evaluation, pollution, cadmium toxicity, stress, toxicity, metabolism, filter feeder community, respiration, nutrient excretion, *in situ* studies

The use of ecosystem or community responses instead of single-species tests in hazard assessment has been emphasized by several authors [*1–4*], together with the importance of detecting sublethal effects, which are visible before more obvious structural changes occur [*5,6*]. Sublethal effects can be measured as changes in functional activities, such as feeding, respiration, primary production, or nutrient excretion [*7*], or as changes in metabolic ratios, such as the oxygen/nitrogen (O/N) ratio, which indicates the relative balance between carbohydrate and protein metabolism and has also been used to describe the physiological status of marine invertebrates [*8–10*].

Usually the community must be enclosed to measure such changes, and this in itself can cause severe experimental problems related to the departure of water quality conditions (for

---

[1]Postgraduate students and docent, respectively, Department of Zoology and Askö Laboratory, Institute of Marine Ecology, University of Stockholm, S-106 91 Stockholm, Sweden.

example, gaseous oxygen) from ambient levels, which thus produces a lack of realism in the results [11]. This problem can be partially solved by using a flow-through system, in which both water and toxicant are continuously replenished and relevant fluxes can be inferred from changes in the inflow and outflow concentrations [12, 13]. Since, in a natural system, fluctuations in light and temperature and variations in the composition of the community among replicates may also conceal responses to disturbance, analytical methods must be used which can distinguish between natural fluctuations and those attributable to treatment. In the past, most researchers relied either on multidimensional state space descriptions of parameter fluctuations [14] or experimental designs in which the communities are monitored before and after treatment and changes are compared with changes in the controls during the same period [15, 16]. In this paper, the authors describe recent improvements made in the design of a portable continuous-flow system for *in situ* bioassay studies, giving an example of its application using one typical littoral zone community exposed to cadmium. Finally, the relative merits of various analytical approaches are discussed, and a new index of disturbance is proposed which appears to be able to identify sublethal stress on communities better than previous approaches.

Methods

In Situ *Bioassay System*

The apparatus (Fig. 1) is a refined version of previously described equipment [13, 15], which has now been adapted for routine bioassay studies. In brief, a submersible pump brings ambient water to an overflow tank, from which it flows through transparent tubes [polyvinyl chloride (PVC), 10 mm in inside diameter] into ten parallel transparent acrylic jars (20 to 30 L in volume) in which the test communities have been placed [13, 15]. A small nozzle (3 mm in inside diameter) creates a current that provides mixing in the jars. Each jar is connected to a solenoid valve, which directs water to outlets or through the measurement chamber. At the outlets, water samples can be taken for nutrient and particle analyses. The measurement chamber contains a stirrer and a polarizing oxygen electrode (Yellow Springs Instrument Inc., Yellow Springs, Ohio), as well as a Pelton flowmeter and a silicon temperature sensor. The height difference between the water surface of the overflow tank and the outlets produces a constant pressure difference that gives stable flow rates through the system, normally adjusted to between 0.5 and 1 L min^{-1}.

The oxygen electrode is provided with a negative polarizing voltage of -0.8 V. The sensor signal is injected into a current-to-voltage amplifier maintaining the electrode anode near ground, which thus prevents interfering current loops through the system [17]. The temperature sensors are provided with 5.00 V from a reference voltage output of the analog-digital (AD) converter. Temperature is also measured inside the electronic equipment to compensate for temperature changes.

The outputs from these signal conditioners and from the light meter (Licor 550B, Lincoln, Nebraska) are connected by means of a twelve-bit analog-digital converter to a microcomputer (Apple II+, Apple Computer Inc., Cupertino, California).

The square-wave pulses from the flowmeter are connected directly to the cassette input of the computer. The solenoid valves are activated by the computer with a simple interface decoding the four-bit output from the game connector to 16 solid-state relays.

To prevent interfering current loops over the mains, the system is connected to a constant-voltage transformer with filters against high voltage peaks. This arrangement also prevents electric shocks since the mains have no galvanic connection to the system. All parts except the computer have normally been run on 12 V DC. The equipment, including ten experimental jars, can be easily transported in an ordinary automobile.

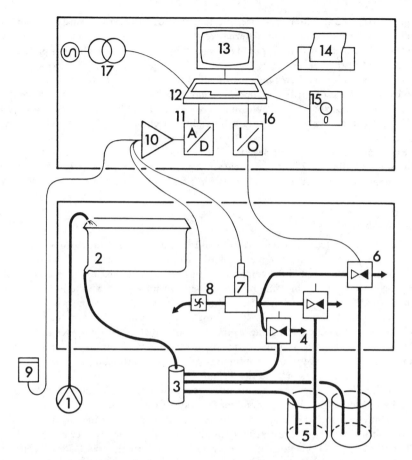

FIG. 1—*Schematic diagram of the* in situ *bioassay system. The lower box represents the* in situ *portion of the system, and the upper box the data recording and storage unit. The components are the submersible pump (1), overflow tank (2), diverter (3), outlet for nutrient sampling (4), experimental jars (5), three-way solenoid valves (6), measurement chamber with oxygen electrode and thermistor (7), flowmeter (8), light meter (9), signal conditioners (10), analog-to-digital converter (11), computer with screen (12, 13), printer (14), disk drive (15), solenoid selector and driver (16), and constant voltage transformer (17).*

Experimental Communities

To minimize disturbances of the littoral zone communities to be tested, these communities are normally placed inside the jars while still attached to their substrates, for example, stones or settling panels where they have been growing. Alternatively, a modified jar can be placed over communities living on rock surfaces or in sediments. Measurements are then made *in situ* for 12 to 24 h before test substances are added in order to obtain values of community parameters before treatment.

Addition of Test Substances

Prior to the addition of test substances to the jars containing littoral communities, the flow is stopped and the test substances are added to the jars, which are then aerated during the exposure time of 10 to 12 h. After exposure, the flow is started again and measurements of the functional parameters are resumed. Alternatively, the test compounds can be added continuously into the flow with a peristaltic pump or by leading the water through tubes painted on the inside with the test substances.

Finally, two or more overflow tanks can be used in which the substances are added to the tanks during on-site effluent testing or when manipulations of temperature and salinity are conducted. In the described example the first protocol was used. The concentrations were 1 and 10 mg L^{-1} of cadmium prepared from solid cadmium chloride (CdCl$_2 \cdot$ 6H$_2$O) dissolved in seawater.

The experiment was performed near the Tjärnö Marine Biological Laboratory on the west coast of Sweden (salinity, 22 parts per thousand; mean water temperature, 18°C). The panels were dominated by the filter feeders *Mytilus edulis* and *Ciona intestinalis*.

Data Processing and Analysis

The computer activates the solenoid valves to direct the outflowing water from the desired jar to the measurement chamber. The means of 100 readings of oxygen, temperature, flow rate, and light intensity are stored in the computer during each 1-min measurement cycle. The next jar is then connected to the measurement chamber. During each cycle the inflowing water is monitored, together with the outflow from one or two jars without organisms, which serve as reference for changes in the characteristics of the inflowing water. Since the same electrode is used for all measurements and the parameters are calculated as the difference between inflow and outflow values, the effects of long-term changes of membrane properties and temperature are compensated for. The results are continuously displayed on the computer screen and on the printer; each 1/2 h they are stored on the diskette.

The difference between the inflow and outflow concentrations of dissolved substances (compensated for residence time in the jars and for flow rate) is used to infer net fluxes of oxygen and nutrients. Daily respiration of algal communities is calculated from the mean of readings in darkness multiplied by 24 h. Gross primary production (GP) is the distance between the respiration baseline and the production peak integrated over a diurnal period. These values are divided by the ash-free dry weight, which is determined at the end of the experiment to obtain weight-specific respiration or production. Single values that are unreasonably low or high (that is, they differ by at least an order of magnitude from the mean) because of electrical interference or air bubbles in the system are discarded. The use of separate oxygen and temperature measurements compensated by the software is a more accurate method than the traditional hardware compensation. The sensitivity is also increased because the same electrode is used to measure the water from all jars. With this arrangement changes of 0.05 mg O$_2$ L^{-1} can be detected. The system can be improved further by the use of a dual electrode system [18] and automatic calibration with aerated water flowing through the measurement chamber after each measurement cycle.

To evaluate the effect of treatment, a perturbation index (PI) is calculated [15]. Each measurement point T_j for the treated jar is divided by the corresponding point of the control jar C_j, and the ratio is calculated by dividing the mean of posttreatment and pretreatment values. The number of samples is n_a before treatment and n_b after treatment. X is oxygen, ammonia, or any other parameter used.

$$PI(X) = \frac{\Sigma \dfrac{T_j}{C_j}}{n_a \Sigma \dfrac{T_i}{C_i n_b}}$$

The perturbation indexes are analyzed statistically by ANOVA [19] for effects of treatment. Scheffé's multiple comparison is used to detect mean differences where appropriate. The standard deviation is also calculated. This does not show in the example graphs because it was generally small (2% of the mean; $n > 200$).

Water samples for determination of nutrients [ammonium (NH_4-N) and phosphate (PO_4-P)] are taken at least two times before and after treatment. The analyses are carried out the same day in the laboratory using standard colorimetric analytical methods [20].

Results

The authors present here briefly the general features of an experiment conducted to test for effects of cadmium exposure on a benthic community consisting mainly of *Mytilus edulis* and *Ciona intestinalis*. The respiration rates ranged from 1.5 to 2.5 mg O_2 g^{-1} h^{-1} before treatment in both the test and control jars; they increased at the 1-mg L^{-1} level but decreased at the 10-mg L^{-1} level of cadmium treatment (Fig. 2a). The untreated jars showed no change in respiration during the course of the experiment. The net NH_4-N and PO_4-P fluxes ranged from 80 to 150 µg nitrogen g^{-1} h^{-1} and from 10 to 50 µg phosphorus g^{-1} h^{-1} (Fig. 2b). During the course of the experiment a general decline in nutrient excretion was observed, even for untreated jars. When these results are expressed as oxygen/nitrogen (O/N) and oxygen/phosphorus (O/P) ratios, they are more constant because the differences in biomass and abundance are eliminated. However, some non-treatment-related variation may remain embedded in these results (for control and treated communities) because of changes in the temperature, salinity, and amount of food present in the inflowing water. When the PI is used, the changes become more apparent, since variation in the parallel controls is normalized and the metabolism of the same communities before treatment is also used as a further control. The PI(O_2) is almost equal to 1.0 for all controls, 1.2 for the low dose and 0.8 for the high dose treatment (Fig. 2c).

The PI(O/N) and PI(O/P) show a greater variation around 1.0 for the controls, but the treated communities consistently have significantly lower indexes, indicating severe disturbance: 0.5 to 0.7 for PI(O/N) and 0.3 to 0.5 for PI(O/P) (Fig. 2c).

Discussion

The use of indexes reduces variation among replicates and treatments due to differences in community structure, biomass, and size distribution. For example, the O/N and O/P ratios will normalize size effects on specific respiration and excretion because size-dependent effects will act similarly on both parameters. The perturbation indexes normalize changes due to temperature, light, salinity, and other environmental conditions since every observation of the treated jar is divided by a corresponding observation of the controls. These general factors affect all jars equally and at the same time (observe the general decline of NH_4-N in Fig. 2b). This is especially evident with primary producers, for which oxygen production varies with solar insolation. Every single point divided by the control value gives a constant value that should ideally be unity (1.0) if the treatments and controls are identical, independent of the net production or consumption of oxygen. Since there are variations in species composition, background metabolism measurements must be made before treatment to establish and compensate for these differences. This is evident in Fig. 2a and b, where the individual controls are significantly different but do not change during the experiment.

The ratios and indexes enable us to distinguish changes due to disturbance by toxic substances, nutrient enhancement, and other artificial factors from natural variations, which are inevitable in real ecosystems. Since the ratios O/P, O/N, and PI are not biomass-related, they allow us to measure the same community repeatedly over long periods of time, even when the biomass changes as a result of growth or death. This is especially useful for continuous monitor-

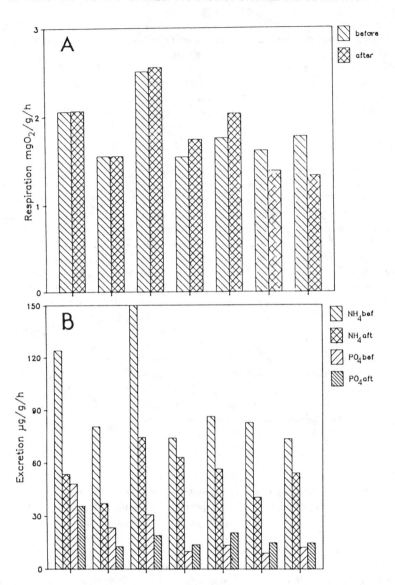

FIG. 2—*Metabolism measurements on settling panel communities in Plexiglas jars exposed to 1 and 10 mg L^{-1} cadmium for 10 h, in comparison with controls [for clarity, the standard deviation (about 2%) is not drawn in the figure]: (a) respiration 12 h before (shaded bars) and 12 h after treatment (cross-hatched bars); (b) PO$_4$-P and NH$_4$-N excretion before and after treatment.*

ing, for example, at a site where pollutants are being discharged or for studying how a community recovers after the disturbance has stopped, because the same microcosm communities are used and their initial and treatment characteristics are already known.

The design of PI used in this study is similar to the BACI design (before and after discharge at both the control and impact sites) proposed by Stewart-Oaten et al. [16]. However, in their application the control and treated sites must be close enough to have the same variation in

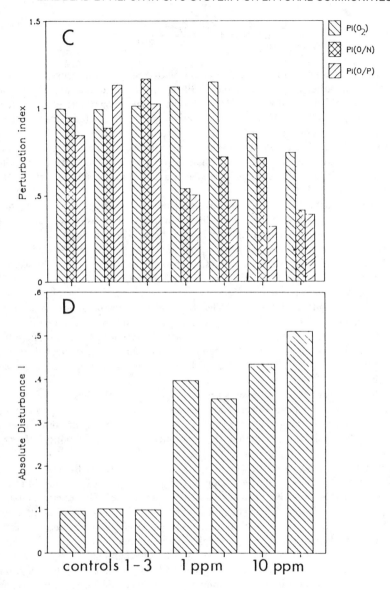

FIG. 2—*Continued:* (c) *perturbation indexes PI(O₂), PI(O/N), and PI(O/P) with the mean value of the controls set to one;* (d) *absolute disturbance index (ADI) calculated from the perturbation indexes given in Fig. 2c.*

light, temperature, and other conditions but be sufficiently separate to avoid disturbance of the control by the treatment. In addition, their method cannot take into account changes in the biomass because they are not looking at exactly the same communities each time.

Even though the variation is reduced by using the PI, there is still variation depending on which parameter is used (Fig. 2*c*). To get a clearer picture of the total effect of the treatment, we propose calculating what we call the absolute disturbance index (ADI), which has the following formula.

$$ADI = \left(\frac{\Sigma[PI(X_i) - 1]^2}{n} \right)^{1/2}$$

This means that all measured parameters are plotted in a multidimensional space, where origin is the undisturbed community and each PI represents one dimension. The absolute distance from origin divided by the number of parameters used, n, summarizes the total effect on all parameters, $PI(X)$. The ADI is the mean of all changes in the system caused by treatment. The ADI differences between treatment and effect values are more evident than the PI differences, as shown in Fig. 2d. The ADI is proportional to the disturbance; larger values indicate a more disturbed situation. The variation in the controls represents the natural variation. This index is always positive, and additional PI are easily added since the square of the ADI value is additive.

The absolute disturbance index originates from Hutchinson's [21] idea of environmental dimensions. In Kersting's [14] description of disturbance, each dimension represents a property of the system in a multidimensional space. He uses the 95% confidence limits to describe the original state space of a two-dimensional system and defines the disturbed situations as points outside this space and the distance from the space as a measurement of disturbance. It is difficult, however, to define confidence limits with several dimensions and to measure the distance from them to a point of disturbance.

Our approach using PI expresses only a change in the measured parameter due to treatment. Since the PI equals one in an undisturbed community, the subtraction of one from the measured PI in an undisturbed community is zero, which means that the points are in origin of the multidimensional space. This eliminates the problem of defining a complicated state space of an undisturbed community. The level of disturbance is then easily measured as the deviation from origin.

Respiration and excretion of nutrients are included in the ADI because they are the most fundamental parts of the metabolism of a community. They affect growth, and thereby also survival, of the community and are very important in the interaction with other communities (for example, algae-animal interactions). In this application oxygen is weighted higher since it is measured more often than nutrients.

The equipment we have described here represents a link between the real ecosystem, the complex mesocosm, and the controlled laboratory experiment and will help us in interpreting observations from effluent recipients and results from laboratories. The functional parameters, for example, respiration, primary production, and nutrient excretion, permit comparable measurements to be made on all hierarchical levels from ecosystems down to single species.

The degree of realism in such a system is high since the communities and ambient factors are close to natural situations, which provides a better basis for accurate predictions of the behavior of the real ecosystem, in contrast to single-species tests in constant laboratory environments. But large model ecosystems with great complexity will often give extremely variable results [22] and are very expensive to replicate and maintain. Using our small systems it is possible to have adequate replicates to obtain statistically reliable results. The perturbation (PI) and absolute disturbance (ADI) indexes reduce the high variations due to differences in insolation, biomass, species composition, and other factors, and make it possible to detect effects of treatment among the noise of natural variation.

Acknowledgments

This study was supported by the Swedish Natural Research Council and the initial parts were also supported by the National Swedish Environment Protection Board. Thanks are due to Prof. Bengt-Owe Jansson, Dr. Walter Boynton, and Dr. Klaus Koop for commenting on the manuscript and to Bibi Mayrhofer for drawing the figures.

References

[*1*] Giddings, J. and Eddlemon, G. K., *Water, Air, and Soil Pollution*, Vol. 9, 1978, pp. 207-212.
[2] O'Neill, R. V. and Reichle, D. E. in *Forests: Fresh Perspectives from Ecosystem Analysis*, R. W. Waring, Ed., Oregon State University Press, Corvallis, OR, 1980, pp. 11-26.
[*3*] Cairns, J., Jr., *Water Research*, Vol. 15, 1981, pp. 941-952.
[4] Cairns, J., Jr., *Environmental Monitoring and Assessment*, Vol. 4, 1984, pp. 259-273.
[5] Perkins, E. J., *Philosophical Transactions of the Royal Society of London, Series B*, Vol. 286, 1979, pp. 425-442.
[6] Matthews, R. A., Buikema, A. L., Cairns, J., Jr., and Rodgers, J. H., Jr., *Water Research*, Vol. 16, 1982, pp. 129-139.
[7] Rapport, D. J., Regier, H. A., and Hutchinson, T. C., *American Naturalist*, Vol. 125, 1985, pp. 617-640.
[8] Bayne, B. L. and Scullard, C., *Journal of the Marine Biological Association of the United Kingdom*, 1977, Vol. 57, pp. 355-369.
[9] Widdows, J., Phelps, D. K., and Galloway, W., *Marine Environmental Research*, Vol. 4, 1981, pp. 181-194.
[*10*] Widdows, J., Donkin, P., Salked, P. N., Cleary, J. J., Lowe, D. M., Evans, S. V., and Thomson, P. E., *Marine Ecology—Progress Series*, Vol. 17, 1984, pp. 33-47.
[*11*] Giesy, J. P., Jr., in *Microcosm in Ecological Research*, *DOE Symposium Series*, No. 52, CONF-781101, National Technical Information Service, Springfield, VA, 1980.
[*12*] Bottom, D. L., *Marine Biology*, Vol. 64, 1981, pp. 251-257.
[*13*] Kautsky, N., *Marine Biology*, Vol. 81, 1984, pp. 47-52.
[*14*] Kersting, K., *Internationale Revue der Gesamten Hydrobiologie*, Vol. 69, 1984, pp. 567-607.
[*15*] Lindblad, C., Kautsky, N., and Kautsky, U., *Ophelia Supplement*, Vol. 4, 1986, pp. 159-165.
[*16*] Stewart-Oaten, A., Murdoch, W. W., and Parker, K. R., *Ecology*, Vol. 67, 1986, pp. 929-940.
[*17*] Forstner, H. in *Polarographic Oxygen Sensors*, E. Gnaiger and H. Forstner, Eds., Springer-Verlag, Berlin, 1983, Chapter I.10, pp. 90-101.
[*18*] Smith, D. F. and Horner, S. M. J., *Marine Biology*, Vol. 72, 1982, pp. 53-60.
[*19*] Orloci, L. and Kenkel, N. C., *Introduction to Data Analysis with Applications in Population and Community Biology*, University of Western Ontario, London, Ontario, Canada, 1984.
[20] Carlberg, S., *New Baltic Manual, International Council for the Exploration of the Sea Cooperative Research Report, Series A*, Vol. 29, 1972, pp. 1-145.
[*21*] Hutchinson, G. E., *Quantitative Biology*, Vol. 22, 1957, pp. 415-427.
[22] Pilson, M. E. Q. and Nixon, S. W. in *Microcosm in Ecological Research*, J. P. Giesy, Jr., Ed., *DOE Symposium Series*, No. 52, CONF-781101, National Technical Information Service, Springfield, VA, 1980, pp. 753-778.

Jan Ahlers,[1] *Erika Rösick,*[2] *and Klaus Stadtlander*[2]

Environmental Chemicals and Biomembranes: Kinetics of Uptake and Influence on Membrane Functions

REFERENCE: Ahlers, J., Rösick, E., and Stadtlander, K., **"Environmental Chemicals and Biomembranes: Kinetics of Uptake and Influence on Membrane Functions,"** *Functional Testing of Aquatic Biota for Estimating Hazards of Chemicals, ASTM STP 988,* J. Cairns, Jr., and J. R. Pratt, Eds., American Society for Testing and Materials, Philadelphia, 1988, pp. 106–119.

ABSTRACT: The uptake kinetics, accumulation, and metabolism of several radioactively labeled environmental chemicals by the eucaryotic microorganism yeast were studied. The results have shown that this test system rapidly provides information about the extent of uptake and accumulation of such chemicals as a function of various external conditions. Thus, different forms of environmental hazard can be simulated.

In addition, in several cases a correlation between inhibition of functional membrane proteins (adenosine triphosphatase and transport systems) and reduction of the cell growth rate was observed, which suggests that the toxicity of an environmental chemical may often be caused by the result of an interaction with important membrane components. The agreement of these results with acute toxicity data from the literature indicate that yeast is suitable for detecting potential hazards to higher organisms from chemicals.

KEY WORDS: hazard evaluation, rapid test system, uptake, accumulation, metabolism, environmental chemicals, functional membrane proteins, nitrilotriacetic acid (NTA), cadmium ion, 4-nitrophenol, hexachlorobenzene, chlorinated phenols

The regulation of toxic substances and their control in various countries makes it necessary for several tests for new chemicals to be compulsory if environmental hazards are to be avoided. These tests include examinations on acute toxicity, mutagenicity, carcinogenicity, and teratogenicity. To induce these effects, the chemical has first to cross the cell membrane to reach its target within the cell. To date, sufficient data concerning the mechanism of uptake by the cell (simple diffusion, mediated transport) and its velocity are available only for a limited number of substances.

For estimating the rate of free diffusion, the octanol/water partition coefficient serves reasonably well. A high partition coefficient would predict a high uptake. However, it has to be kept in mind that the composition of a natural membrane is much more complex than can be described by the simple octanol/water partition coefficient [1,2]. Moreover, biological membranes consist largely of proteins, which have a strong influence on the fluidity of the surrounding lipid domains.

As the plasma membrane is a most important part of the cell, containing several enzymes and transport systems which supply the cell with nutrients and with energy, it is also of interest to check the possible interference of environmental chemicals with these cell functions. Such a reaction could well be the reason for toxic effects on the cell or the organism, especially as the

---

[1]Federal Environmental Agency, D-1000 Berlin 33, Federal Republic of Germany.
[2]Institut für Biochemie und Molekularbiologie, Freie Universität Berlin, D-1000 Berlin-33, Federal Republic of Germany.

most hazardous chemicals possess a high octanol/water partition coefficient and thus accumulate in the hydrophobic part of the membrane.

The legal regulation and control of toxic substances requires a large amount of information about the more than 100 000 compounds produced to date by industry. It is therefore desirable to establish test systems using microorganisms or mammalian cell cultures to gain knowledge about the ecotoxicity of all these substances. Such test systems could be used to reduce the number of experiments with higher organisms.

According to guidelines by the Organization for Economic Cooperation and Development (OECD), such a test system should be generally applicable, reproducible, standardizable, and inexpensive [3]. These requirements can be easily met by using microorganisms. Eucaryots are more advantageous than procaryots since their more complex structure allows a better extrapolation of the results to higher organisms. Another advantage of microorganisms is that it is possible to work with them in closed systems so that even volatile substances can be examined.

In this contribution, the authors chose a number of substances with quite different structures and toxicities and examined the rates and mechanisms of uptake by yeast cells as well as the factors of accumulation. In order to gain information on whether the toxicity of these compounds to living cells may be due to their reaction with functional membrane proteins, we studied their influence on the plasma membrane H^+-adenosine triphosphatase (ATPase) and on the glucose as well as purine transport systems. We compared the results with data obtained from physiological investigations on cell growth performed with the same substances and with data existing in the literature on acute toxicity of these agents to higher organisms. If the effects on cell growth and on functional membrane proteins are in the same range, the toxic effects of these chemicals may well be due to their interaction with the cell membrane.

A positive result of our experiments could then be that they provide a basis for establishing a new test system which would reduce the number of experiments with higher animals that are normally necessary in studies of potentially ecotoxic substances.

Material and Methods

Reference Chemicals

The following reference chemicals (of the highest purity available) were used:

NTA	=	nitrilotriacetic acid (Merck, Darmstadt, Federal Republic of Germany [FRG])
$CdCl_2$	=	cadmium chloride (Riedel-de-Haen, Seelze, FRG)
4-NP	=	4-nitrophenol (Fluka, Buchs, Switzerland)
PCP	=	pentachlorophenol (Riedel-de-Haen, Seelze, FRG)
TCP	=	2,4,6-trichlorophenol (Merck, Darmstadt, FRG)
2,4- and 2,6-DCP	=	dichlorophenol (Aldrich, Steinheim, FRG)
4-CP	=	4-chlorophenol (Merck, Darmstadt, FRG)
HCB	=	hexachlorobenzene (Riedel-de-Haen, Seelze, FRG)

The radioactive labeled reference chemicals nitrilotri(1-^{14}C)acetic acid, [115m]CdCl$_2$, hexachloro(U-^{14}C)benzene, and 4-nitro(2,6-^{14}C)phenol were products of Amersham Buchler (Braunschweig, FRG); [14]C-(U)-pentachlorophenol was obtained from the Commissariat à l'Energie Atomique (Gif-sur-Yvette, France). The specific activities for NTA, HCB, and 4-NP were in the range of 1.5 to 4 megabecquerel (MBq)/μmol and for cadmium 0.74 to 4.4 MBq/mg.

Growth Conditions

The yeast *Saccharomyces cerevisiae* was grown under aerobic conditions at 30°C in a medium containing 2% glucose, 1% Difco yeast extract, and 5% peptone and was harvested in the stationary growth phase for all experiments, unless described otherwise. In some cases, as indicated in the text, the cells were pretreated for 1 h with 2% (weight/volume) glucose in 50 mM citrate at pH 5 under aeration to achieve maximal uptake activity [4].

To examine the kinetics of growth, the reference chemicals were either added to the yeast cells in the logarithmic phase, or, in order to examine the effect on the lag phase, 5×10^6 stationary cells per millilitre were added to the growth medium, which was supplemented with various concentrations of the chemical. The increase in cell density was checked by reading the optical density at 530 nm. A calibration curve relating the optical density to the number of cells per millilitre was established. All the values are averages of four or more experiments.

Uptake Experiments

For the uptake experiments, 2×10^6 to 4×10^7 cells per millilitre were incubated under continuous shaking at 25°C with the radioactive substrate in 10 or 50 mM 2-(N-morpholino)ethane sulfonic acid (MES). The uptake was determined at pH 5 for NTA and HCB and at pH 6 for Cd^{2+}, PCP, and 4-NP (the latter also at pH 8 for comparison). Aliquots of 1 mL were collected at the indicated times on glass fiber filters (GF 92, Schleicher and Schüll, Dassel, FRG) previously soaked with stopping solution, containing the unlabeled chemical. The filters were washed with 30 mL of this solution and were then assayed for radioactivity in a Beckman scintillation counter (LS 7000).

The experiments were performed in triplicate. The filtration conditions were varied for each chemical to ensure that the background radioactivity on the filter did not exceed 5% of the total uptake activity of the cells.

A differentiation between unspecific binding and intracellular accumulation was performed in the cases of NTA, Cd^{2+}, and PCP by experiments with dead cells (incubated at 70°C for 60 min, which resulted in a cell viability of less than 0.005%) and, in the case of Cd^{2+}, by experiments with cells that did not transport Cd^{2+} because of the absence of glucose [5].

The problem of unspecific adsorption of HCB and 4-NP will be discussed later in the paper in connection with the interpretation of the uptake versus time plots.

The internal cell concentrations of the reference chemicals were calculated, assuming a volume of 1.2×10^{-13} L/cell [6]. The dry weight of a cell was 3×10^{-8} mg.

Metabolic Conversion Studies

To determine the intracellular degradation and conversion products of the reference chemicals, the cells were extracted either by the lyophilization procedure of Nazar et al. [7], using an ice-cold solution of 1 M acetic acid in ethanol, or by the method of Folch et al. [8] and Bligh and Dyer [9] with chloroform and methanol after homogenizing the cells with glass beads. The extractable portions of radioactivity were 85% and 93%, respectively, for the two methods. The cell-free extracts were differentiated by thin-layer chromatography (TLC), and the radioactive spots were detected by autoradiography. The solvent system for NTA was n-butanol/pyridine/water (3:3:5), for PCP it was n-hexane/acetone/acetic acid (70:30:2), and finally, for 4-NP it was ethyl acetate.

Measurement of ATPase Activity

Plasma membranes from *Saccharomyces cerevisiae* were obtained from protoplasts by the procedure described previously [10]. The adenosine triphosphatase (ATPase) activity was determined by continuously recording the amount of inorganic phosphate released [11]. The incubation time was 10 min. Continuous recording of the product formed during the reaction enabled us to detect possible deviations from steady-state conditions. The assay under standard conditions contained 50 mM MES, titrated to pH 6.0 with Trisbuffer, 0.5 mM magnesium chloride

(MgCl$_2$), 0.5 mM adenosine triphosphate (ATP), 50 mM potassium chloride (KCl), and 20 to 50 μg of membrane protein in a total volume of 5 mL. Various concentrations of these compounds were added 2 min before starting the reaction with ATP. All the values are averages of four or more experiments performed in duplicate.

Measurement of Purine and Glucose Transport

For purine uptake experiments, cells were pretreated in 50 mM citrate buffer at pH 3.5 or 5.5 containing 2% glucose, for 1 h to ensure maximum purine (^{14}C) uptake activity of the cells [4]. The initial uptake velocity of 8-^{14}C-hypoxanthine (2 MBq/μmol) was determined as described by Forêt et al. [12].

For glucose uptake, stationary cells were directly taken after being washed with the uptake buffer. The slightly modified procedure of Bisson and Fraenkel [13], using D-(^{14}C(U))-glucose (12.5 MBq/μmol), was used. The uptake time was 1 min. The uptake rate was constant for the first 5 min.

For the inhibition studies, the cells were incubated with several concentrations of the compounds mentioned for 15 min before the reaction was started. The experiments were done at least three t¹mes in duplicate.

Results and Discussion

Uptake Experiments

NTA, a substance probably not very toxic, was chosen because it is under discussion as a substitute for polyphosphates in detergents and thus will probably be found in relatively large quantities in wastewater.

To estimate the initial rates of uptake and the factors of accumulation, the increase in internal concentration of NTA in the cell with time was plotted under various experimental conditions at a concentration of 10 mM in the external medium (data not shown). It was apparent that the uptake of NTA by cells from the logarithmic phase was rather slow. Even after 120 min an equilibrium was not achieved.

Glucose preincubation led to a large increase in uptake and, after 120 min saturation was almost reached. Furthermore, the accumulation factor was much higher in this case. The uptake was even more enhanced when cells from the stationary phase were chosen for the experiments. Probably the function of the plasma membrane as a permeability barrier was reduced when the cells reached the stationary phase. Preincubation of the cells with glucose seemed to have a similar effect. Table 1 summarizes the results deduced from these curves.

In the presence of glucose, NTA is possibly metabolized more rapidly, leading to an additional enhancement of incorporation of radioactivity into the cell. We observed an almost complete metabolizing of NTA within 2 h since there is a significant decrease in intensity of the NTA spot, with a retardation factor (Rf) of 0.86, in comparison with two more polar spots with Rf = 0.77 and Rf = 0.83 (not shown).

A substrate variation over a wide concentration range can be used to distinguish between the two basic mechanisms of uptake: simple diffusion and catalyzed transport. By plotting the initial uptake rate versus the substrate concentration, saturation kinetics should be observed in the case of catalyzed transport, whereas for simple diffusion, the rate should be directly proportional to the concentration gradient across the membrane. When performing the experiments with logarithmic, or stationary, cells in the presence or absence of glucose, we observed a linear relationship between the initial rate of uptake and the NTA concentration (0.1 to 100 mM NTA), clearly indicating that this chemical enters the cell by free diffusion [6].

Cd^{2+} was chosen as the reference ion because of the abundant presence of heavy metal ions in the environment and because of their many hazardous effects described in the literature

TABLE 1—*Initial rates of NTA uptake and accumulation factors after 2 h of uptake.*[a]

Glucose Preincubation[b]	Growth Phase	NTA Uptake, $\left(\dfrac{\mu\text{mol/g dry weight} \times \text{min}}{\text{mol/L}}\right)$	Accumulation Factor	
			Based on Wet Weight, $\text{NTA}_{\text{internal}}/\text{NTA}_{\text{external}}$	Based on Dry Weight, $\left(\dfrac{\mu\text{mol/g dry weight}}{\mu\text{mol/mL medium}}\right)$
−	logarithmic	10	0.3	1.2
+	logarithmic	467	2.4	9.8
−	stationary	1330	3.7	14.7
+	stationary	1800	3.9	15.6

[a]The values are the means of three experiments.
[b]The minus sign indicates no preincubation; the plus sign indicates preincubation.

[14–16]. The uptake of the radioactive isotope cadmium-115m was rapid and largely dependent on incubation conditions. The initial uptake rate, in nanomoles per minute times the dry weight in milligrams, was 1.2 in the presence of glucose and 0.03 in its absence. The factors of accumulation were 2120 in the presence and 80 in the absence of glucose (calculated on the dry weight basis). Uncouplers such as dinitrophenol or ATPase inhibitors such as azide led to a strong inhibition of uptake [17]. These observations suggest that the metal ions are taken up by an energy-dependent transport system. In agreement with this assumption, we observed saturation kinetics when plotting the initial rates versus the external concentration (Fig. 1). The reciprocal plot of these data resulted in a straight line (inset in Fig. 1) from which the Michaelis constant, K_m, and the maximal uptake rate, V_{max}, of the transport system, were calculated: $K_m = 0.3 \pm 0.1$ mM and $V_{max} = 5.2 \pm 2.4$ nmol per minute and milligram of dry weight. The presence of glucose is necessary for full activity.

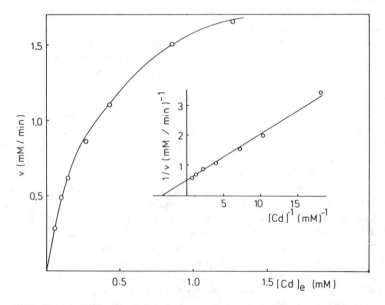

FIG. 1—*Influence of the external cadmium concentration on the initial uptake rate in the presence of glucose. The inset shows the double reciprocal plot of cadmium uptake.*

As cadmium is not an essential metal ion, but rather toxic to living cells, it is probably not transported by a specific cadmium carrier but uses a transport system that has evolved for essential trace metal ions such as zinc; this system has been described by Fuhrmann and Rothstein [18]. As this transport system was not formed during evolution for the uptake of cadmium, it obviously has a lower affinity for this ion than for, for example, zinc. A similar interpretation is given by Norris and Kelly [14], who observed a K_m of 1 mM for cadmium uptake by yeast.

Substituted phenol or benzene derivatives are known to be very hazardous to all kinds of organisms. Therefore, the study of their effects was predominant in this investigation. In contrast to NTA and cadmium, they possess a relatively high octanol/water partition coefficient.

Figures 2 through 4 show the time course of HCB, 4-NP, and PCP uptake, obtained with cells from both the logarithmic and stationary growth phases. In the case of HCB, because of its low solubility, experiments could only be performed with concentrations up to 5 μg/L. We observed an extremely rapid incorporation during the first few seconds of uptake, possibly caused by unspecific adsorption at the outer face of the cells (Fig. 2). However, the high lipophilicity of

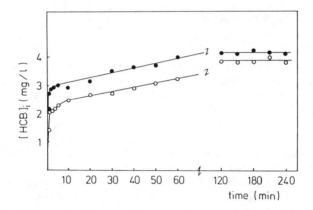

FIG. 2—*Intracellular concentration of HCB as a function of time: (○) logarithmic cells, (●) stationary cells. The HCB concentration was 5 μg/L (1 mg/L HCB = 3.5 μmol/L).*

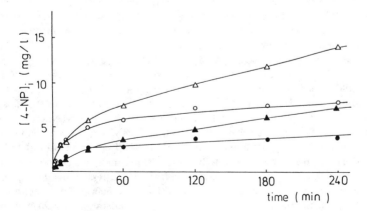

FIG. 3—*Intracellular concentration of 4-NP as a function of time: (△) logarithmic cells, pH 6; (○) stationary cells, pH 6; (▲) logarithmic cells, pH 8; (●) stationary cells, pH 8. The 4-NP concentration was 70 mg/L (1 mg/L 4-NP = 7.2 μmol/L).*

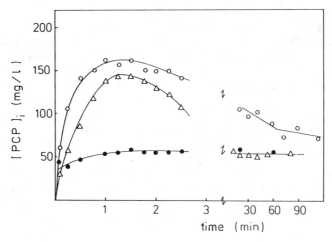

FIG. 4—*Intracellular concentration of PCP as a function of time: (○) logarithmic cells, (△) stationary cells, (●) dead cells. The PCP concentration was 0.5 mg/L (1 mg/L PCP = 3.8 μmol/L).*

this substance makes it probable that the large initial uptake is caused by accumulation in the hydrophobic regions of the plasma membrane. During the following 60 min HCB probably diffused slowly into the cytoplasm and was additionally accumulated in the inner membrane system.

The uptake curves for 4-NP (Fig. 3) are similar to those of HCB, indicating a biphasic uptake at both pH 6 and pH 8. As for HCB, the rapid uptake in the initial seconds could be due to an accumulation in the plasma membrane, followed by a slow transport into the cytoplasm. It is apparent that the undissociated form of 4-NP led not only to a higher initial uptake rate but also to a higher uptake capacity after 4 h. In contrast to HCB, the accumulation potential of 4-NP was much lower, and an intracellular concentration equivalent to that in the external medium could not be achieved. Because of the low uptake rate, considerably higher concentrations of 4-NP, in comparison with HCB, had been applied in this set of experiments.

Figure 4 shows the uptake of PCP by cells of both growth phases and, in addition, by heat-inactivated cells. Within 2 min of uptake, living cells attained a maximum of accumulation, followed by an efflux of radioactivity. Dead cells, however, showed a lower accumulation capacity and maintained this level. It is noteworthy that, after the efflux of radioactivity, the living cells equilibrated at the same concentration as the dead cells. Control experiments revealed that the efflux could not be explained by a loss of viability.

Substrate variations showed that the uptake of 4-NP, PCP, and HCB occurred by passive diffusion, which was rapid in the cases of HCB and PCP but much slower with 4-NP (not shown). In Table 2, the initial rates of uptake and the accumulation factors of these three compounds are summarized. These data were compared with those obtained for the alga *Chlorella* by Geyer et al. [19]. These authors revealed a positive correlation between the octanol/water partition coefficient, P_{ow}, and the accumulation factor, f, for 45 organic compounds, expressed by the regression equation $\log f = 0.681 \times \log P_{ow} + 0.863$. In the case of HCB and PCP, our values fit this regression equation very well; however, the accumulation factors of 4-NP in yeast deviate by approximately two orders of magnitude from the regression line for algae. Possible reasons for this deviation are the different test conditions and the use of different organisms as test systems [20]. Furthermore, it should be taken into consideration that 24 h of incubation with 4-NP are probably not sufficient for complete equilibration, since at this time the intracellular concentration had not reached that of the medium. This low accumulation, measured by the amount of ^{14}C present in the cells in comparison with the medium, could not be due to a

TABLE 2—*Initial rates of uptake and accumulation factors[a] for 4-NP, PCP, and HCB as measured with cells from the stationary growth phase.*

Chemical	Log P_{ow}[a]	pH	Initial Uptake Rate, $\left(\dfrac{\mu mol/g \text{ dry weight} \times min}{mol/L}\right)$	Accumulation Factor,[b] $\left(\dfrac{\mu mol/g \text{ dry weight}}{\mu mol/mL}\right)$
4-NP	1.9	6	1.16×10^2	0.77
		8	4.33	0.25
PCP	3.7	6	1.93×10^6	6.45×10^2
HCB	5.5	5	3.73×10^6	1.6×10^4

[a] The *n*-octanol/water partition coefficient.
[b] Accumulation factors on the basis of dry weight were calculated by the quotient of the concentration in the cells and the final concentration in the external medium after 2 h of uptake for HCB and PCP and after 24 h for 4-NP.

transport of 4-NP conversion products out of the yeast cells since no such product was detected by TLC analysis.

The metabolic conversion studies for PCP demonstrated that no PCP derivatives passed into the medium within 2 h, since only the spot of PCP itself could be detected in the medium by TLC. This suggests that the efflux of PCP is possibly due to a damaged membrane structure, although there was no decrease in the viability of the cells. However, the analysis of the cell-free extract (aqueous phase) revealed that it was a fraction more polar than the PCP itself, which remained at the origin of the TLC sheet, indicating that PCP conjugates could have been formed as described by Langebartels and Harms [21].

Influence on Membrane Functions

In order to find out whether the toxic effect of a chemical can be related to its interaction with functional membrane proteins, we studied the influence of the reference chemicals on the plasma membrane ATPase, as well as on the purine transport system and the glucose carrier. We compared the results with data obtained from physiological investigations on the kinetics of cell growth performed with the same substances.

Neither NTA nor its magnesium or calcium complexes had any influence on cell growth up to millimolar concentrations. Only NTA at above a 5-mM concentration led to a reduction of cell growth, probably because of complexation of essential metal ions within the cell. After a few hours of incubation, this effect disappeared because NTA is metabolized and the bound metal ions are released.

In accordance with these results, neither ATPase nor the transport systems studied were influenced by up to millimolar concentrations of NTA and its calcium or magnesium complexes.

Investigating the influence of cadmium ions on the ATPase, we observed a strong dependence of inhibition on the pH and magnesium concentration [5]. From dose-response curves under favorable conditions (that is, at pH 8 and 1 mM Mg^{2+}) and at 0.2 mM MgATP, we calculated a lowest effective concentration of Cd^{2+} at $< 1 \mu M$ and 50% inhibition at 3.5 μM.

In contrast to the ATPase, the transport systems investigated in this study were inhibited by much higher concentrations of Cd^{2+}. This different effect of cadmium on proteins of the same membrane rules out the possibility that the influence of cadmium is due to an unspecific alteration of the membrane fluidity, especially since we could show that the purine transport system is as influenced by alterations of membrane fluidity as is the ATPase [1].

In addition, we were interested in finding out whether these data correlate with physiological effects of cadmium on yeast cells. We therefore examined the kinetics of cell growth in the presence of up to 50 μM CdCl$_2$. We observed inhibition of cell growth with increasing concentrations of cadmium. The lowest effective concentration was 2 μM (230 $\mu g/L$), whereas at 50 μM (6 mg/L) the cells ceased growth completely. The median effective concentration (EC$_{50}$) was 13 μM (1.5 mg/L). These values are in the lower range of data reviewed by Trevors et al. [16].

As we had observed a high initial uptake rate of HCB and a large factor of accumulation, we expected an influence on cell growth too. However, we could not detect any influence either on growth or on functional membrane proteins. We assume that the low solubility of HCB in water (5 $\mu g/L$), its high volatility, and unspecific adsorption effects prevent the accumulation of considerable amounts within the cell. The relatively high median lethal dose (LD$_{50}$) values on acute toxicity in the literature [22] are in agreement with this assumption.

The last group of substances studied was the substituted phenols. Besides the question of whether a correlation exists between an influence on growth rate and an influence on functional membrane proteins, we wanted to find out if a correlation can be found between structure and reactivity for this class of chemicals and, thus, if it is possible to deduce from a given structure of a chemical its hazardous effect.

The influence of these phenols on growth kinetics was in most cases similar to the results obtained for PCP (Fig. 5). We observed a decrease in growth rate during the logarithmic phase as well as a reduction of cell density in the stationary phase. However, when the experiments were started with starved cells, a slight increase in the lag phase could also be observed (not shown). These results suggest that the chemical affects the cell by reacting with more than one site, for example, DNA and enzymes. The lowest effective concentration calculated from the dose-response curve was 10 μM. Figure 6 presents the dose-response curves of several chlorine-substituted phenols on cell growth (for PCP see Fig. 7). It can be seen that there are large differences in effectiveness, depending on the number of chlorine atoms.

As the substituted phenols possess a high octanol/water partition coefficient, it can be expected that they accumulate in the membrane and interact with functional membrane proteins. We observed, indeed, an inhibition of both ATPase and the transport systems studied at rather low concentrations of PCP. The effect occurred in the same range where inhibition of growth rate was observed (Fig. 7) and was pH dependent (not shown). At lower pH a stronger inhibition could be observed, indicating that the undissociated form of the chemical is effective.

Besides PCP, all the substituted phenols under investigation (TCP, 2,4-DCP, 2,6-DCP, 4-CP and, 4-NP) inhibited the ATPase as well as the transport systems. The mechanism of inhibition seems to be different for each functional membrane protein, since the inhibition of, for exam-

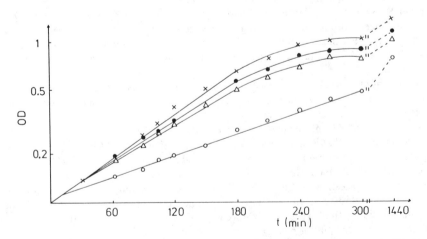

FIG. 5—*Kinetics of cell growth in the presence of varying amounts of PCP: (×) control, (●) 4 mg/L, (△) 6 mg/L, (○) 10 mg/L.*

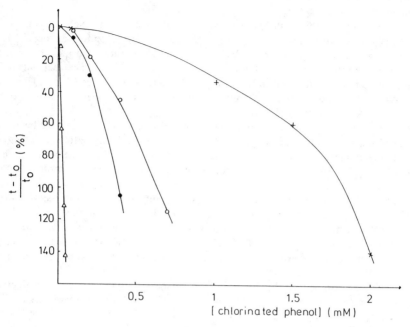

FIG. 6—*Influence of chlorine-substituted phenols on cell growth: (×) 4-Cl phenol, (●) 2,4-Cl₂ phenol,* *(○) 2,6-Cl₂ phenol, (△) 2,4,6-Cl₃ phenol. The symbol t indicates the generation time in the presence or absence (t₀) of chlorinated phenol.*

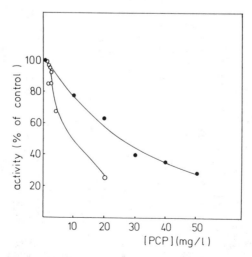

FIG. 7—*Effects of PCP on growth (○) and ATPase activity (●). The conditions for the ATPase measurements at pH 8 were Mg²⁺, 1 mM; MgATP, 0.2 mM. The ATPase activity in the absence of PCP was 1.6 units per milligram of protein.*

ple, 4-NP was due to a decrease in the V_{max} of the purine transport system and also due to a combined effect on K_m and V_{max} of the glucose carrier (Fig. 8).

Figure 9 summarizes the results obtained with several substituted phenols. It is apparent that there is a significant correlation between the concentration of the chemical leading to 10% inhibition of growth and the lipophilicity. Phenols with a high octanol/water partition coefficient are more potent inhibitors of growth than those with a lower P_{ow}. A linear dependency could also be observed by plotting the concentration leading to 10% inhibition, I_{10}, of ATPase activity versus log P_{ow}, which suggests that there is a close relationship between effects on growth and effects on the ATPase activity (Fig. 9).

Our results show that the factors of accumulation obtained do not fit very well with the P_{ow} values, which are generally taken to predict the potential of bioaccumulation of a chemical. As expected, the main exception are cadmium ions, which are taken up by the cell through a very efficient transport system and can be bound to sulfhydryl (SH) groups of proteins within the cells [23].

In contrast, we observed for 4-NP a much lower factor of accumulation, as should be expected from its log P_{ow}. Obviously, the plasma membrane functions in this case as a high-permeability barrier, preventing equilibration between the inside and outside concentration even after 24 h of incubation. These results indicate that it is preferable to measure the bioaccumulation, using eucaryotic microorganisms, than to determine the octanol/water partition coefficient.

From Table 3 it can be seen that the P_{ow} values do not correspond in all cases with effects on cell growth when different classes of substances are compared. However, there is a good correlation between the effects of the chemicals on cell growth and their effects on functional membrane proteins. The results fit well with data on acute toxicity to fish, as well as to mammalian cell cultures, and reasonably with data on acute toxicity to rats, obtained from the literature [22,24-26]. These observations indicate that the eucaryotic microorganism yeast is also suitable for predicting the hazard potentials of chemicals for higher organisms.

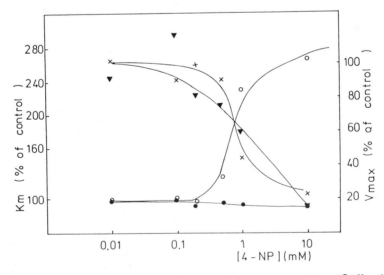

FIG. 8—*Effect of 4-NP on the* K_m *and* V_{max} *of both transport systems:* ($\blacktriangledown$) V_{max}, ($\bullet$) K_m *of the purine transport system, pH 3.5;* ($\times$) V_{max}, ($\bigcirc$) K_m *of the glucose transport system, pH 5. Control experiments showed the following values (100%). For hypoxanthine transport:* $V_{max} = 54 \pm 10 \times 10^{-18}$ *mol/cell* $\times$ *min* ($2 \pm 0.3 \times 10^{-9}$ *mol/min* $\times$ *mg dry weight);* $K_m = 1.4 \pm 0.2$ *μM. For glucose uptake:* $V_{max} = 3.2 \pm 0.4 \times 10^{-15}$ *mol/cell* $\times$ *min* ($107 \pm 13 \times 10^{-9}$ *mol/min* $\times$ *mg dry weight);* $K_m = 2.3 \pm 0.5$ *mM. These results are in agreement with data in literature [4,13].*

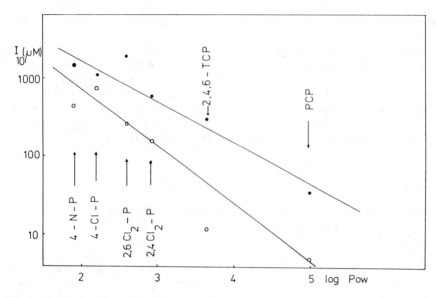

FIG. 9—*Correlation between the lowest effective concentration, I_{10}, of substituted phenols on the cell growth (○) and ATPase activity (●) and log P_{ow}. The following phenols were tested: 4-NP, 4-Cl phenol, 2,4-DCP, 2,6-DCP, 2,4,6-TCP, and PCP.*

TABLE 3—*Summary of the results in this study.*

Chemical	I_{10}, mg/L[a]				
	Hypoxanthine Transport System	Glucose Transport System	ATPase	Yeast Cell Growth	Log P_{ow}
NTA	>3000	>3000	only effects due to complexing properties	<3000	<0
Cd^{2+}	1.2	550	0.11	0.23	<0
4-NP	74	45	180	50	1.9
PCP	0.4	0.6	4	2	3.7
HCB	>0.01	>0.01	>0.01	>0.01	5.5

[a]I_{10} = concentration of the chemical leading to 10% inhibition, considered to be the lowest effective concentration.

When it is possible to obtain the chemicals radioactively labeled, this system can be easily used to measure the initial uptake rate. With these results, it is possible to clarify the mechanism of uptake. Moreover, one can predict to what extent a chemical will be taken up under certain environmental conditions, including pH, substrate concentrations, and the presence of other effectors (for example, the influence of a complexing agent such as NTA on cadmium uptake [17]). Continuous monitoring of the accumulation of chemicals by cells is advantageous not only in comparison with determination of the octanol/water partition coefficient but also in comparison with discontinuous measurement of the bioaccumulation after, for example, 24 h. It is easily possible to subtract the amount of unspecific adsorption to the cell. In addition, metabolizing of the substance can be detected by this system and can thus be taken into account.

Because of their high lipophilicity, many environmental chemicals are accumulated in the hydrophobic region of the membrane, where many enzyme and transport systems, most important for the function of the whole cell, are located. Thus, the functional proteins in the membrane can be inhibited by rather low concentrations of environmental chemicals in the external medium. These effects may play an important part in the overall toxicity of hydrophobic xenobiotics without there necessarily being a reaction of the chemical with intracellular compounds or compartments such as the nucleus, mitochondria, or cytoplasm. Such effects can be easily detected with this system.

Conclusions

1. Using the eucaryotic microorganism yeast, it was possible to measure the rate and mechanism of uptake of several environmental chemicals. Moreover, the extent of bioaccumulation as well as of metabolism was examined.

2. Functional tests with the same chemicals showed that similar concentrations of environmental chemicals inhibited growth and functional membrane proteins. These observations support the view that, in many cases, the toxic effect of a substance may be due to its interaction with cellular membranes.

3. The test systems described are easy to handle, reproducible, and fast. The results agree well with data obtained using mammalian cell cultures and fish, which suggests that predictions of aquatic toxicity are possible.

4. The octanol/water partition coefficient does not correlate in all cases with the toxicity of a substance. Within one class of substances—in this study phenol derivatives—there was a linear relationship allowing structure-reactivity predictions. In contrast, the toxic effects of chemicals taken up by the cell via transport systems, such as cadmium, cannot be predicted by their lipophilicity. Also, substances with very low solubility in water, such as hexachlorobenzene, do not react as would be expected from their octanol/water partition coefficients.

Acknowledgment

This study was supported by a grant from the German Federal Environmental Agency, Berlin, FRG.

References

[1] Stadtlander, K., Rade, S., and Ahlers, J., *Journal of Cellular Biochemistry*, Vol. 20, 1982, pp. 369-380.
[2] Stein, W. D. in *Membrane Transport*, S. L. Bonting, and J. J. H. de Pont, Eds., Elsevier, Amsterdam, 1981, pp. 1-28.
[3] Klein, A. W., Harnisch, M., Poremski, H. J., and Schmidt-Bleek, F., *Chemosphere*, Vol. 10, 1981, pp. 153-207.
[4] Reichert, U. and Winter, M., *Biochimica et Biophysica Acta*, Vol. 356, 1974, pp. 108-116.
[5] Ahlers, J. and Rösick, E., *Toxicological and Environmental Chemistry*, Vol. 11, 1986, pp. 291-300.
[6] Rösick, E., Stadtlander, K., and Ahlers, J., *Chemosphere*, Vol. 14, 1985, pp. 529-544.
[7] Nazar, R. N., Lawford, H. G., and Tze-Fei-Wong, J., *Analytical Biochemistry*, Vol. 35, 1970, pp. 305-313.
[8] Folch, J., Lees, M., and Sloane-Stanley, G. H., *Journal of Biological Chemistry*, Vol. 226, 1957, pp. 497-509.
[9] Bligh, E. G. and Dyer, W. J., *Canadian Journal of Biochemistry and Physiology*, Vol. 37, 1959, pp. 911-917.
[10] Ahlers, J., *Canadian Journal of Biochemistry*, Vol. 62, 1984, pp. 998-1005.
[11] Arnold, A., Wolf, H. U., Ackermann, B. P., and Bader, H., *Analytical Biochemistry*, Vol. 71, 1976, pp. 209-213.

[12] Forêt, M., Schmidt, R., and Reichert, U., *European Journal of Biochemistry*, Vol. 82, 1978, pp. 33-43.

[13] Bisson, L. F. and Fraenkel, D. G., *Proceedings of the National Academy of Sciences of the USA*, Vol. 80, 1983, pp. 1730-1734.

[14] Norris, P. R. and Kelly, D. P., *Journal of General Microbiology*, Vol. 99, 1977, pp. 317-324.

[15] Ord, M. J. and Al-Atia, G. R., *The Chemistry, Biochemistry and Biology of Cadmium*, M. Webb, Ed., Elsevier/North-Holland Biomedical Press, New York, 1979, pp. 141-173.

[16] Trevors, J. T., Stratton, G. W., and Gadd, G. M., *Canadian Journal of Microbiology*, Vol. 32, 1986, pp. 447-464.

[17] Ahlers, J. and Rösick, E., *Bulletin of Environmental Contamination and Toxicology*, Vol. 37, 1986, pp. 96-105.

[18] Fuhrmann, G. F. and Rothstein, A., *Biochimica et Biophysica Acta*, Vol. 163, 1968, pp. 325-330.

[19] Geyer, H., Politzki, G., and Freitag, D., *Chemosphere*, Vol. 13, 1984, pp. 269-284.

[20] Ellgehausen, H., Johann, A. G., and Esser, H. O., *Ecotoxicology and Environmental Safety*, Vol. 4, 1980, pp. 134-157.

[21] Langebartels, C. and Harms, H., *Zeitschrift für Pflanzenphysiologie*, Vol. 113, 1984, pp. 201-211.

[22] Rippen, G., *Handbuch der Umweltchemikalien*, 2nd ed., Ecomed, Landsberg, Germany, 1987.

[23] Vallee, B. L. and Ulmer, D. D., *Annual Review of Biochemistry*, Vol. 41, 1972, pp. 91-128.

[24] "Collection of Minimum Pre-Marketing Sets of Data Including Environmental Residue Data on Existing Chemicals," Umweltbundesamt, Berlin, 1982.

[25] Ahlers, J., Benzing, M., Gies, A., Pauli, W., and Rösick, E., *Chemosphere*, Vol. 17, 1988, pp. 1603-1615.

[26] Foret, M. and Ahlers, J., *Journal of Ecotoxicology and Environmental Safety*, Vol. 16, in press.

Eric L. Morgan,[1] *James R. Wright, Jr.,*[2] *and Richard C. Young*[3]

Developing a Portable Automated Biomonitoring System for Aquatic Hazard Evaluation

REFERENCE: Morgan, E. L., Wright, J. R., Jr., and Young, R. C., **"Developing a Portable Automated Biomonitoring System for Aquatic Hazard Evaluation,"** *Functional Testing of Aquatic Biota for Estimating Hazards of Chemicals, ASTM STP 988,* J. Cairns, Jr., and J. R. Pratt, Eds., American Society for Testing and Materials, Philadelphia, 1988, pp. 120-127.

ABSTRACT: In the spring of 1986, an intensive aquatic biological testing program was carried out over a 45-day period in an attempt to evaluate possible toxicity in an important fishery resource in the French Broad River of East Tennessee. Since 1969 a marked decline in sauger *(Stizostedion canadense)* populations has occurred below Douglas Dam. In addition to toxicity tests with the water flea (*Ceriodaphnia* spp.), fathead minnow *(Pimephales promelas),* and sauger in various life stages, a portable automated fish breathing rate biomonitoring system (AFIRMS) was field tested at streamside in a mobile bioassay trailer. The personal computer based AFIRMS was equipped with 24 channels for continuously monitoring adult sauger breathing activities and various physical sensors. Though all the tests revealed no apparent toxicity in the tailwater reaches of the river, a simulated field test with AFIRMS using toxic leachate from a landfill source induced immediate sauger responses and ultimate mortality to most of the test animals.

KEY WORDS: automated aquatic biomonitoring, functional biological testing, hazard evaluation, toxicity tests, mobile bioassay trailer

In 1970, Cairns et al. [1] proposed a biological monitoring system for watershed drainages that would provide an early warning of water pollution. In addition to stream surveys and biomonitoring at critical areas throughout the drainage, Cairns and his co-workers described a unique system for automatically recording fish breathing and swimming activities in response to developing toxicity in effluents and ambient receiving waters. Support for this approach has been based on the observation that a variety of aquatic animals produce discrete bioelectric action potentials as a result of specific physiological activities. For example, the intimate contact between the water and the respiratory surfaces of fish gills, and, presumably, gills of other aquatic animals, makes these organs particularly susceptible to toxic materials [2]. Changes in rhythmic patterns can be detected electronically [3] and the information managed by inexpensive computer-assisted systems [4-8].

Automated aquatic biomonitoring systems in operation today typically are custom designed and fabricated to meet only a few specific requirements, for example, recording fish breathing rates or selected behavior activities and physical water quality sensor outputs [9-12]. Limitations imposed by highly specialized automated systems have restricted their use in more generalized streamside applications. In an attempt to overcome these problems, the authors of this

[1]Associate professor, Department of Biology, Tennessee Technological University, Cookeville, TN 38505.
[2]Manager, Special Projects and Research, Division of Air and Water Resources, U.S. Tennessee Valley Authority, Knoxville, TN 37901.
[3]President, Young-Morgan & Associates, Inc., Franklin, TN 37064.

paper are at present testing portable automated monitoring devices that have the capacity to monitor various biological functions from a number of different aquatic animals continuously, while simultaneously recording any number of physical sensor inputs. In addition to the sensor versatility, D-C-powered portable systems have the advantage of various data communication options from remote locations, including satellite relay systems and modem transfer [13–15]. Using a similar portable system, complemented by a personal computer for data management and supported by various physical water quality sensors and an event-triggered water sampler, preliminary tests have been completed from a mobile bioassay trailer at streamside [16]. Our purpose here is to discuss current technical developments and applications of these portable automated biomonitoring systems as complements to functional biological testing in hazard evaluations.

Portable Automated Biomonitoring Systems

Streamside Mobile Trailer Facilities

A mobile bioassay trailer [17] was located at a streamside station about 39 km below Douglas Dam on the French Broad River (FBR) in East Tennessee for a 45-day test period during the spring of 1986. Auxiliary facilities accompanying the trailer consisted of A-C power hookups, pumps and polyvinyl chloride (PVC) piping for water supplies, hold tanks, a portable 14-m³ environmental chamber, and a fiberglass tank mounted on a truck bed for transporting reference water to the site from nearby Little Pigeon River.

Adjacent to the trailer, the Styrofoam-insulated environmental locker was positioned at streamside to receive water from the gravity supply head tanks (Fig. 1). Two water supply systems were set up—one delivering recirculating-replenishment reference water and one administering once-through water pumped from the adjacent FBR. The gravity flow lines and pump lines were constructed of PVC plumbing, while the head tanks were commercially available fiberglass structures.

Automated Fish Biomonitoring System

The major components of the automated fish biomonitoring system (AFIRMS) were divided into two subsystems: a mobile trailer, which housed most of the electrical instrumentation, and a portable walk-in environmental locker for maintaining the biosensing chambers for sauger (*Stizostedion canadense*) (Fig. 1). Inside the locker, two groups of eight 38-L Styrofoam aquaria with covers were positioned to receive pass-through flows from head tanks. The water supply systems for both groups of eight aquaria were designed so that the flows of reference and river water could be mixed or delivered singly. The flow to each aquarium was maintained at approximately 500 mL/min. On the inner surface of each Styrofoam cover, a small yellow-red light-emitting diode (LED) lamp was installed to provide a 12-h light : 12-h dark photoperiod.

Positioned inside each aquarium was a biosensing chamber similar in design to those used in previous studies [13–15,17]. The chambers were constructed of lengths of 20-cm-diameter PVC pipe, allowing clearance of several centimetres from the inside walls of the aquaria. Attached to the inner surface of each chamber was a pair of parallel-probe-type antennae, and a sauger was positioned between the probes inside each chamber and then isolated in a separate aquarium (Fig. 1). Once completed, a total of 16 sauger (204 to 449 g) were simultaneously maintained under reference water flows of 500 mL/min for an acclimation period of 9 days. On the second day of acclimation, the sauger were subjected to a 2-h exposure of 15 ppm Terramycin.

In the trailer, the discrete sauger ventilatory (breathing) events were received through coaxial cables and amplified up to 5×10^5 times before being digitized for temporary storage in the buffers of the interface board located in the desk-top computer. Using the AFIRMS-basic pro-

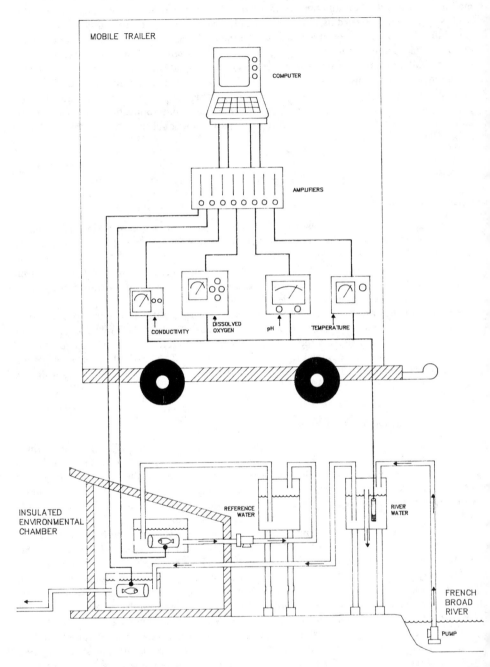

FIG. 1—*Schematic of the portable automated aquatic biomonitoring system operating from a mobile bioassay trailer at streamside.*

gram, buffered data in the interfaced unit could be requested at a rate from 1 to 59 min, as selected by the operator. In this study, a 4-min monitoring cycle was used when data were requested—a 3-min recording interval and a 1-min wait period. The data were logged in an appropriate format and filed on hard disks. Accompanying the breathing rate data were simultaneously recorded physical measurements for temperature, pH, dissolved oxygen, and conductivity.

Hazard Assessment

Following the 9-day period of acclimation to reference water flows, one group of 8 sauger was subjected to FBR water at a rate of 500 mL/min, while the remaining group of 8 continued to receive reference flows. Approximately half the 1.3-m^3 volume of the recirculating reference water supply was replenished daily. Over the remaining 35 days of the study, during which the sauger breathing rates were continuously recorded, a complement of aquatic toxicity tests was performed in the laboratory with water samples taken at five locations along the FBR reach below Douglas Dam. The tests included fathead minnow *(Pimephales promelas)* larval survival and growth studies and *Ceriodaphnia* spp. survival and reproduction tests [16]. Attempts were made to assess the ambient river toxicity using sauger egg-hatching and larval-survival studies; however, because of complications from disease, made worse by algae entanglement, definitive testing was postponed until 1987.

Simulated Toxic Stress Test

After more than 40 days of field testing with AFIRMS, operating under ambient river water conditions, the 12 sauger remaining in the test chambers were subjected to a stepwise treatment of approximately 0.9 and 1.9% toxic leachate collected from a landfill. In previous tests, this leachate was found to cause significant mortality to *Ceriodaphnia* and fathead minnows in static-renewal laboratory toxicity evaluations [16].

Prior to initiating the AFIRMS leachate treatments, the once-through FBR flow water was converted to a recirculating system similar to the reference water system. To accomplish the test treatments, 35 L of leachate was initially added to 402 L of recirculating FBR water, and approximately 3 h later the concentration was increased to 1.9%. The flow rates to the sauger were maintained at about 500 mL/min.

Toxics Evaluation

Aquatic Toxicity Assays

No apparent toxicity was observed from FBR water to sauger in the field tests or to *Ceriodaphnia* and fathead minnows in the laboratory during the period of study. There were no significant differences found between five FBR sources and a Tennessee River control in the survival or mean dry weights of larval fathead minnows tested for 7 days. From the 100% concentrations tested, the average survival of fathead larvae ranged from 87% for the Little Pigeon River reference water to 93% for several FBR sites. Similar results were found for *Ceriodaphnia*, with no significant difference observed in survival between the control and treatment groups over the 7-day test period. However, reproductive success was substantially improved in most of the FBR treatments in comparison with the controls and reference groups (Table 1). The water chemistry values taken during testing were well within the tolerance ranges for *Ceriodaphnia* and fatheads.

TABLE 1—*Summary of results from the Fathead Minnow (*Pimephales promelas*) larval survival and growth test and from the* Ceriodaphnia *spp. survival and reproduction test for the French Broad River toxicity study.*[a]

	Fathead Minnow		Ceriodaphnia	
Water Source[b]	Survival, %	Dry Weight, mg	Survival, %	Young, No.
Control				
M	100	0.528	100	15.4
TR	...	...	90	13.1
Station 1 (FBR, Douglas Dam), %				
100	93	0.421	100	12.9
75 + 25 M	97	0.527	100	17.1
75 + 25 TR	...	...	100	14.1
Station 2 (FBR, Mile 26.5), %				
100	93	0.535	100	24.3[c]
75 + 25 M	93	0.550	100	29.4[c]
75 + 25 TR	...	...	100	27.8[c]
Station 3 (LPR, Mile 1.0), %				
100	93	0.519	100	27.4[c]
75 + 25 M	87	0.510	90	29.3[c]
75 + 25 TR	...	...	100	33.9[c]
Station 4 (FBR, Mile 10), %				
100	87	0.547	100	18.6
75 + 25 M	90	0.517	100	33.7[c]
75 + 25 TR	...	...	100	35.8[c]
Station 5 (LPR, Mile 10.5), %				
100	87	0.480	100	15.8
75 + 25 M	97	0.470	100	21.3
75 + 25 TR	...	...	90	16.4

[a]Each *Ceriodaphnia* value is the average of 10 replicates, and each fathead minnow value is based on 3 replicates of 10 fish each (1986).
[b]Key to abbreviations:
   M = culture medium dilution water.
   TR = Tennessee River dilution water.
   FBR = French Broad River water.
   LPR = Little Pigeon River water.
[c]Significantly different from the culture medium controls at the 90% level of confidence.

AFIRMS Field Test

Partial analysis and evaluation of the AFIRMS test shows that a total of 34 adult sauger were monitored throughout the study period. Of this number, 21 were tested in reference and 13 in FBR water for the 45-day field test. The high replacement and turnover of sauger in the reference group was attributed primarily to gravity flow interruptions. This disruption, essentially a clogging effect, was caused by filamentous algae which developed in the recirculating reference water head tank.

Breathing rates for 9 sauger surviving the duration of the field test were compared (Fig. 2). Average rates for the 7-day interval of the acclimation period, when all the fish received reference flows, were compared with rates recorded for the balance of the study (35 days), when one group of sauger continued receiving reference water while the other group received FBR flows. Although not significant, the average breathing rates had a tendency to be lower during this 35-day period for 3 of 5 reference fish and all 4 FBR exposed sauger. In general, the average rates were lower and more consistent for the FBR-exposed group than for the reference group.

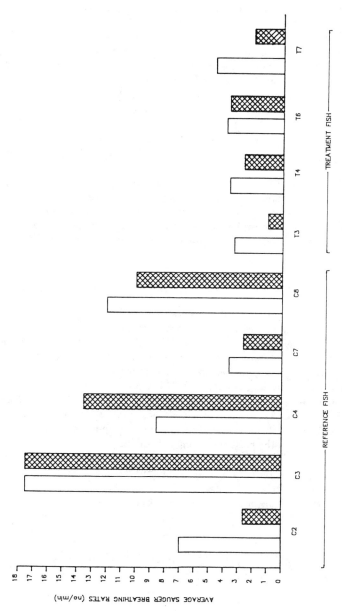

FIG. 2—*Mean breaths per minute for sauger recorded from a mobile trailer with the automated fish respiration monitoring system. Comparisons are made between a 7-day acclimation period (open bars) and the test period (cross-hatched bars) for 5 reference and 4 treatment fish. For acclimation all the fish received reference water. During the test period the reference fish continued receiving reference water, while the treatment fish were subjected to water from the French Broad River (spring 1986).*

The greater differences and overall increased breathing rates among the reference fish may have been the result of reduced flows and a concomitant temperature increase that resulted in the transported reference water head tank. The temperature varied by as much as 5 degrees Celsius from morning to evening. The minimum values were 7.5 and 5.5°C for the FBR and reference waters, respectively; the highest temperatures reached were 20 and 21°C. The reference water values were higher for 52 measurements, lower for 6, and the same as the FBR values for 7 readings. The largest difference between the FBR and reference waters was 2.5 degrees Celsius, with an average of 0.8 degrees. These differences were found to be significant (paired t-test) at the 0.0001 probability level. Other complicating factors may have been differences in the water chemistry. The lower buffering capacity of the reference water, with its characteristic low hardness and conductivity, could have imposed subtle physiological stresses on the sauger.

Though the results provide encouragement in support of AFIRMS application to hazard assessment, these preliminary evaluations are not sufficient to permit a final determination of sauger stress responses to exposure to French Broad River waters. The reduced breathing rates recorded for FBR-exposed sauger may, in effect, be a reflection of a stress-related response. Typically, stress-related breathing responses in fish have been associated with erratic and increased breathing rates [9,10,12].

The toxic leachate stress tests revealed that the material was acutely toxic to *Ceriodaphnia* and fathead minnows at concentrations less than 1%. At streamside, when approximately 1.9% leachate was delivered to 12 sauger being monitored by AFIRMS, immediate increases in breathing rates resulted. Within several hours of exposure, mortalities occurred, and by 13 h exposure, 11 of the 12 sauger displayed no breathing responses and were found to be dead on examination.

Summary

Evaluations of toxic materials carried out with streamside AFIRMS tests and ones done in the laboratory using *Ceriodaphnia* and fathead minnows revealed little or no toxicity in French Broad River waters below Douglas Dam during the spring of 1986. These results may be biased since stream flows were unusually low because of a record drought for this period. If toxic compounds were present in runoff flows following storm events in the drainage, this potential contribution to the evaluation of toxics was not assessed. The automated fish respiration monitoring system was demonstrated as a real-time portable biomonitoring technique for field applications. Apart from problems associated with algae-clogged water supply systems, the automated biomonitor performed well and the results support future applications in mobile trailers. With minor modifications of the portable automated aquatic biomonitor, one may expect to see future applications in water supply systems of permanent space stations [18] or possibly in early warning systems for earthquake events. In a similar application to stream acid deposition evaluations [7], in which a portable stand-alone AFIRMS is linked to satellite communications [15], watershed networking for real-time biological monitoring has been realized.

Acknowledgments

Support for this study was provided by the U.S. Tennessee Valley Authority and by the authors. We gratefully appreciate the assistance of Helen Stoops.

References

[1] Cairns, J., Jr., Sparks, R. E., and Walters, W. T., "Biological Systems as Pollution Monitor." *Research and Development*, Vol. 9, No. 70, 1970, pp. 22-24.
[2] Schaumburg, F. D., Howard, T. E., and Walden, C. C., *Water Research*, Vol. 1, 1967, pp. 731-737.
[3] Camougis, G., *Turtox News*, Vol. 38, 1960, pp. 156-167.

[4] Bonner, W. P., and Morgan, E. L., "On-Line Surveillance of Industrial Effluents Employing Chemical Physical Methods and Fish as Sensors," Technical Report No. B-030TN, Tennessee Water Resources Center, University of Tennessee, Knoxville, TN, 1976.

[5] Cairns, J., Jr., and Gruber, D., *Bioscience*, Vol. 29, No. 11, 1979, p. 665.

[6] Morgan, E. L., Herrmann, R., Eagleson, K. W., and McCullough, N.D., *Water International*, Vol. 5, 1980, pp. 23-27.

[7] Morgan, E. L., Young, R. C., and Crane, C., "Automated Multi-Species Biomonitoring Employing Fish and Aquatic Invertebrates of Various Trophic Levels," *Freshwater Biological Monitoring*, D. Pascoe and B. W. Edwards, Eds., Pergamon Press, Oxford, England, 1984.

[8] Morgan, E. L., Young, R. C., Crane, C. N., Armitage, B. J., "Developing Automated Multispecies Biosensing for Contaminant Detection," *Water and Science Technology*, Vol. 19, No. 11, 1987, pp. 73-84.

S. Dennis, Eds., American Society for Testing and Materials, Philadelphia, 1986.

[9] Cairns, J., Jr., and van Der Schalie, W. H., "Early Warning Biological Monitoring," *Biological Monitoring in Water Pollution*, Pergamon Press, New York, 1982.

[10] Doane, T. R., Cairns, J., Jr., and Buikema, A. L., Jr., "Comparison of Biomonitoring Techniques for Evaluating Effects of Jet Fuel on Bluegill Sunfish *(Lepomis macrochirus),*" *Freshwater Biological Monitoring*, D. Pascoe and B. W. Edwards, Eds., Pergamon Press, Oxford, England, 1984, pp. 103-112.

[11] Henriksen, A., Rogeberg, E., Anderson, S., and Veidel, A., "Mobil Lab Niva—A Complete Station for Monitoring Water Quality," Report 11, Norwegian Institute for Water Research, Oslo, Norway, 1986, p. 44.

[12] Morgan, W. S. G., and Kuhn, P. C., "A Method to Monitor the Effects of Toxicants upon Breathing Rates of Largemouth Bass (*Micropterus salmoides* Lacepede), *Water Research*, Vol. 8, pp. 767-771.

[13] Morgan, E. L., Eagleson, K. W., Herrmann, R., and McCullough, N. D., *Journal of Hydrology*, Vol. 51, 1981, pp. 339-345.

[14] Morgan, E. L., and Eagleson, K. W., "Remote Biosensing Employing Fish as Real-Time Monitors of Water Quality Events," *Hydrological Applications of Remote Sensing and Remote Data Transmission*, Publication No. 145, International Committee on Remote Sensing, Washington, DC, 1983, pp. 235-243.

[15] Morgan, E. L., Eagleson, K. W., Weaver, T. P., and Isom, B. G., "Linking Automated Biomonitoring to Remote Computer Platforms with Satellite Data Retrieval in Acidified Streams," *Impact of Acid Rain and Deposition on Aquatic Biological Systems, ASTM STP 928*, B. G. Isom, S. Dennis, and J. Bates, Eds., American Society for Testing and Materials, Philadelphia, 1986, pp. 91-94.

[16] Tomljanovich, D. A., Wright, J. R., Jr., Schweinforth, R. L., Seawell, W. M., Rodes, A. H., McDonough, T. A., and Morgan, E. L., "A Study to Investigate Potential for Toxicity for the French Broad River to Sauger," U.S. Tennessee Valley Authority, Office of Natural Resources and Community Development, Knoxville, TN, 1987, p. 126.

[17] Morgan, E. L., and Young, R. C., "Automated Fish Respiratory Monitoring Systems for Stream Acidification Episode Assessment," *Verhandlungen International Vereinigung Limnology*, Vol. 22, 1984, pp. 1432-1435.

[18] Morgan, E. L., Young, R. C., Smith, M. D., and Eagleson, K. W., "Rapid Toxicity Detection in Water Quality Control Utilizing Automated Multi-Species Biomonitoring for Permanent Space Stations," *Journal of Environmental Sciences*, Vol. 30, No. 2, 1987, pp. 47-49.

Donald C. Mcnaught,[1] Scott D. Bridgham,[2] and Craig Meadows[1]

Effects of Complex Effluents from the River Raisin on Zooplankton Grazing in Lake Erie

REFERENCE: Mcnaught, D. C., Bridgham, S. D., and Meadows, C., **"Effects of Complex Effluents from the River Raisin on Zooplankton Grazing in Lake Erie,"** *Functional Testing of Aquatic Biota for Estimating Hazards of Chemicals, ASTM STP 988,* J. Cairns, Jr., and J. R. Pratt, Eds., American Society for Testing and Materials, Philadelphia, 1988, pp. 128–137.

ABSTRACT: Functional ecosystem tests should reflect the hazards of toxic chemicals, as well as stimulation by nutrients, by measuring a single flux of phytoplankton to the dominant members of the community. The flux of phytoplankton and detritus to zooplankton is reflected by the filtering rates of individual organisms, expressed as millilitres per animal per hour. The authors used common particle counting techniques to measure such fluxes in the waters of Lake Erie. They then examined the impact of complex effluents on the filtering rates. These effluent effects are scored as inhibition or stimulation of filtering by the dominant herbivores in the Lake Erie ecosystem. In the River Raisin, a tributary to Lake Erie, specific effluents usually inhibited grazing by the herbivores *Daphnia, Diaptomus,* and *Cyclops,* although one effluent was stimulatory. These results were directionally consistent and probably depended on the characteristics (especially the concentrations of metals) of the effluents. The inhibitions were also of considerable magnitude. The authors recommend the use of such a secondary ecosystem functional bioassay (SEFB) for detecting hazards in point sources or tributaries to large systems, such as Lake Erie.

KEY WORDS: hazard evaluation, functional testing, filtering rate, zooplankton, herbivores, ecosystems, inhibition, stimulation, River Raisin, Lake Erie

Great Lakes planktonic systems, like many other aquatic systems, have been exposed to contaminants for decades. Initial attempts to understand, avoid, and ultimately correct former contamination have resulted in the development of separate functional and structural approaches to measuring sensitive parameters. A forthcoming review will examine the successes and failures of functional versus structural testing with respect to hazards in the Great Lakes Basin [1]. Evans and McNaught [1] have concluded that tests involving single species exposed to complex effluents allow evaluation of the toxicity of real-world waters. Furthermore, they concluded that the most useful functional tests have been based on characteristics of zooplankton reproduction, quantified as net reproductive rates, R_0, or the population growth rate, r. Few studies have measured the effects of pollutants on herbivore ingestion. It would seem that ingestion is a logical functional indicator of environmental hazards, because decreased grazing may result in reduced fecundity of these herbivores. The filtering rates (in millilitres per animal per hour), when multiplied by the food biomass (in milligrams per millilitre), provide a direct measurement of the ingestion rate (in milligrams per animal per hour) and, when used alone, provide an indirect measurement.

[1]Professor and technician, respectively, Department of Ecology and Behavioral Biology, University of Minnesota, Minneapolis, MN 55455.
[2]Graduate student, School of Forestry and Environmental Studies, Duke University, Durham, NC 27706.

Grazing has been used to quantify the hazards associated with single contaminants and, in rare instances, with complex effluents. Mirza [2] showed that grazing by *Daphnia magna* was reduced by polluted waters. Cooley [3] found that filtering by *Daphnia retrocurva* was reduced when animals were exposed to pulp mill effluents. In central Lake Huron, the grazing rates of herbivores were normal but became depressed an average of 87% in polluted Saginaw Bay [4]. All these examples use grazing to estimate the presence of an environmental pollutant, but none identifies the hazard. When looking at the effects of single isomers of common contaminants, with exposures at realistic environmental levels, tests utilizing measurements of grazing have likewise been successful. A pure isomer, dichlorobiphenyl, had no effect on grazing by adult *Cyclops* and nauplii in Lake Huron [4]; however, the degradation product of this compound reduced grazing by 25%. In a similar fashion, toxaphene reduced grazing in Lake Superior by both *Diaptomus sicilis* (8 to 65%) and *Limnocalanus macrurus* (3 to 57%), depending on exposure levels [5]. Similarly, in laboratory tests a single contaminant (lindane) reduced grazing by herbivores [6]. These studies are easier to interpret than those using mixed effluents, but in ecosystems impacted by man it is necessary to understand the impact of complex effluents, because the Great Lakes, along with other large ecosystems, are impacted in this way.

In this volume on functional hazard testing, the authors of this paper have decided to focus on results of functional tests employing species grazing rates because this test alone was the most effective in identifying environmental hazards in a detailed study of the River Raisin in southern Michigan. Hazard tests employing effects of complex effluents on gross primary productivity, bacterial uptake of acetate, and zooplankton reproduction were less successful, when compared side by side using the same effluents.

Study Area

The study area in the lower River Raisin flows through a heavily industralized complex. It is described in detail elsewhere in this volume. We include a map of the lower river (Fig. 1) to emphasize the relationship of the three important stations where effluents were collected: namely, the Monroe Wastewater Treatment Plant (WWTP), denoted Station 7, Mason Run (Station 8), and the discharge from the Ford Motor Co. (Station 9).

Experimental Procedure for Estimating Inhibition of Grazing

Herbivores (rotifers, cladocerans, and copepods) are the principal converters of phytoplankton into animal protein for utilization by fishes. Contaminants such as polychlorinated biphenyl compounds (PCBs) have been demonstrated to have significantly reduced grazing [7]. The filtering rates by zooplankton on the natural algal community of the Lake Erie ecosystem were used to determine the inhibition of this vital ecosystem function. Inhibition or stimulation was studied using dilutions of complex effluents.

Phytoplankton were collected from Lake Erie and grown in the laboratory in a culture chamber (Percival 1-60). These phytoplankton were cultured so that their numbers were about equal to those of the natural environment, although the composition undoubtedly changed. For example, in determinations of filtering by *Diaptomus*, the cell numbers decreased over the time of the experiment from 2.5% (September) to 10.3% (July). The density of specific particles in the environment at the onset of these experiments was 454 to 4776 cells per millilitre [in the range of 3 to 40 μm in equivalent spherical diameter (ESD)].

Dilutions of phytoplankton cultures were made with water from the river/lake series of stations, assuming that it was contaminated. Controls were made using lake waters.

The zooplankton species included those forms dominant in Lake Erie (*Diaptomus* spp., *Cyclops bicuspidatus*, and *Daphnia* spp.). The animals were used only in the months they were abundant. All animals were acclimated for 4 h to phytoplankton foods before the introduction

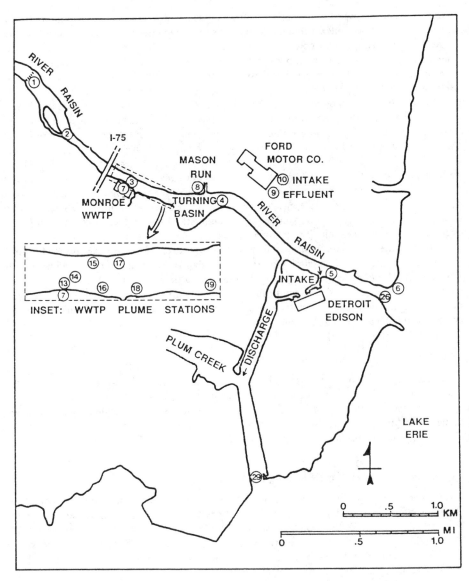

FIG. 1—*The sampling stations in Monroe Harbor on Lake Erie and the River Raisin 1983 and 1984.*

of contaminated or "control" water. Adult animals (ten of a single species) were transferred into experimental feeding vessels (with a food volume of 100 mL) in carriers (tubing with netting to allow the passage of water and food).

Zooplankton grazing, expressed as the filtering rate in these functional estimates, is most accurately measured in the laboratory with a particle counter [8]. Calculation of the filtering rates using a presorted size category was done by employing the equation of Gauld [9], which requires information on the density of particulate foods, both before and after the introduced animals have fed for 15 min. The filtering rates (in millilitres per animal per hour) represent the volume of water from which particles (within the size category) were removed, produced, or both. The filtering rates were expressed as a function of the equivalent spherical diameter

(ESD) and as a mean for the arbitrary range of 3 to 40 μm ESD. This range of sizes has been used by other investigators as it represents the size of food normally consumed by zooplankton in natural situations.

Results

Functional testing for hazardous substances requires a test or series of measurements that are repeatable with relatively low variability and that represent the impact of contaminants or stimulants on a significant segment of an aquatic system.

Initially, we selected four tests to examine impacts on trophic levels below the level of fishes. All four tests measured single functional responses [10] that we commonly understood as major fluxes. These included the bacterial uptake, phytoplankton gross photosynthesis, zooplankton grazing, and zooplankton reproduction (two measurements). The results from the functional tests involving gross photosynthesis are reported elsewhere in this volume (the paper by Bridgham et al.).

The test for inhibition or stimulation of zooplankton grazing proved to be the most valuable of the four tests. A detailed analysis for one month's data is presented showing predominantly inhibition of grazing by complex effluents from three outfalls/streamwaters on the River Raisin (Table 1). These data are for a month (September 1983) when major grazers were present in the natural system, including *Daphnia retrocurva* and *Diaptomus* spp. A summary is presented for 1983 (Figs. 2a through 2c), showing the consistent relationship between grazing impacted by complex effluents from the WWTP (Station 7), expressed as filtering rates in effluents and control waters from Lake Erie. For one of the months graphed (July), the third major grazer, *Cyclops bicuspidatus*, was present.

Raw data (Table 1) show changes in grazing by these important herbivores, but the responses are varied. Complex effluents from WWTP (Station 7) and Mason Run (Station 8) inhibited grazing by the large calanoid *Diaptomus* and the cladoceran *Daphnia retrocurva*. Grazing by the calanoid was inhibited 65 to 90%, and by the cladoceran 24 to 71%. Neither organism showed increased inhibition with increased dosage. Effluents from the Ford Motor Co. discharge (Station 9) stimulated grazing 14 to 18%. The impact of these effluents on *Diaptomus* is not shown, because in this one case the controls did not feed.

Monthly summaries for an effluent and its impact on grazing show similar changes (Figs. 2a through 2c). For most dates, except July 1983, sample dilutions from the Monroe WWTP (Station 7) caused an inhibition of zooplankton filtering. For September, filtering by *Diaptomus* and *Daphnia retrocurva* was inhibited in relation to the control filtering; in October, filtering by *Daphnia schodleri* was inhibited; and for April and May 1984, filtering by *Diaptomus* was inhibited. In July 1983, filtering by *Cyclops* and *Diaptomus* was stimulated (Figs. 2a and 2c).

The reasons for inhibition of secondary ecosystem function by complex effluents from Monroe WWTP are not totally clear. DiToro et al. [11] examined the relationship between the zooplankton filtering rate and free residual chlorine and found that the filtering rate increased with the residual chlorine concentration. Furthermore, when filtering was depressed, these events were correlated with high concentrations of zinc and copper [11]. Filtering was not compared with concentrations of organic contaminants, which must be considered, since PCBs and toxaphene inhibited grazing [4,5].

The filtering rates were likewise inhibited by effluents from Mason Run (Station 8). *Diaptomus* showed inhibited feeding during July and September 1983 and April and May 1984. Most often (July, September, and May) filtering decreased with increased dosage.

For the Ford Motor Co. discharge (Station 9), the filtering rates increased with the dose, showing stimulation. At lower concentrations (10 and 25%), the filtering was equivalent to the control values.

It is possible that inhibition of filtering by herbivores due to contaminants is common in the spring and fall but that stimulation occurs during warmer months. Only one example of this

TABLE 1—Filtering rates and changes relative to controls for two major herbivores exposed to 10, 25, and 50% additions of complex effluents from three sources on 16 Sept. 1983.

| | Source and Concentration of Complex Effluent | | | | | | | | |
| | WWTP (Station 7) | | | Mason Run (Station 8) | | | Ford Motor Co. Discharge (Station 9) | | |
Herbivore	10%	25%	50%	10%	25%	50%	10%	25%	50%
Diaptomus spp., mL/animal/h (% change relative to controls)	0.14 (−71)	0.16 (−67)	0.17 (−65)	0.03 (−90)	0.03 (−90)	0.03 (−90)	...	...	...
Daphnia retrocurva, mL/animal/h (% change relative to controls)	0.38 (−71)	0.29 (−29)	0.20 (−57)	0.36 (−67)	0.29 (−24)	0.22 (−42)	1.16 (+14)	1.20 (+18)	1.20 (+18)

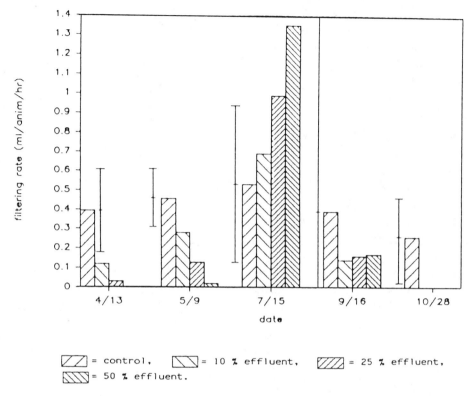

= control,   = 10 % effluent,   = 25 % effluent,
= 50 % effluent.

FIG. 2a—*Secondary ecosystem functional bioassay (SEFB) for Monroe WWTP effluents for* Diaptomus.
The range in controls is indicated by the vertical bars.

phenomenon was available. In July *Diaptomus* was stimulated by all the effluent dilutions and stimulation increased with the dose, whereas in April, May, September, and October, inhibition increased with the dose (Fig. 3).

Discussion

The secondary ecosystem functional bioassay (SEFB) was used in tandem with three other bioassays (involving bacteria, phytoplankton, and zooplankton). It was the only bioassay with limited variability so that hazards presumably related to the occurrence of contaminants could be detected in complex effluents entering a large aquatic ecosystem. Side-by-side comparisons were made with the Mount-Norberg *Ceriodaphnia* bioassay [12,13]. *Ceriodaphnia* life tables were run both for 7 days, as prescribed, and for 40 days or longer to determine the net reproductive rate for each parthenogenic female. The results of these functional tests showed too much variability to detect hazardous substances in River Raisin waters. For the *Ceriodaphnia* test, the standard error (SE) was 19.5% of the mean reproductive rate, R_0. For the gross primary productivity bioassay, the standard error was 44% of the mean rate of primary productivity. For the uptake of acetate by heterotrophs, the standard error was 18% of the mean rate of uptake (Table 2). Thus the SEFB has certain advantages in that it has been demonstrated under field and laboratory conditions to produce consistent answers on specific complex effluents. The SEFB also exhibited some serious problems and was characterized by some costly but not difficult aspects.

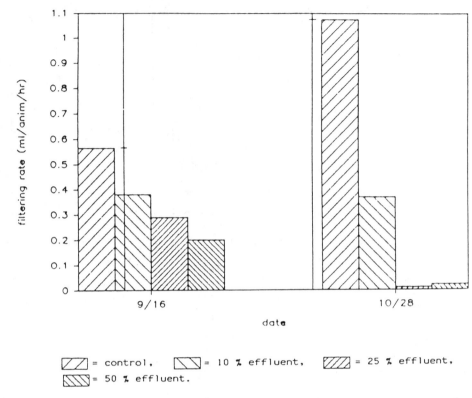

FIG. 2b—*Secondary ecosystem functional bioassay (SEFB) for Monroe WWTP effluents using* Daphnia. *The range in controls is indicated by the vertical bars.*

The most serious problem is the necessity of storing effluents for testing. Effluents should be collected daily, because their toxicity changes in storage,[3] at least with respect to *Ceriodaphnia*. Since locally occurring control waters, with their natural algal assemblages are required, air shipment and maintenance of control water in lighted, temperature-controlled cabinets is a necessity. Control waters and effluents were stored for between 8 days (July) and 33 days (September). This storage may not be as detrimental as thought, because the filtration rate of *Diaptomus* was only depressed slightly. In the May experiment the filtering was depressed 0.5% per day of storage, and in April 0.6% per day. Likewise, natural assemblages of grazers are used. *Diaptomus* and *Cyclops* do not readily reproduce in the laboratory, and adults were collected at the same time as control water samples were taken. *Daphnia* can of course be readily cultured by a variety of methods. All forms must be maintained on native foods or be acclimated to natural assemblages of algae before testing.

Conclusions

1. Zooplankton filtering rates constitute a sensitive and consistent bioassay for complex effluents. The filtering rates respond to an increased dosage of effluents by exhibiting either stimulation or inhibition. This bioassay is called the secondary ecosystem functional bioassay (SEFB).

[3]D. I. Mount, U.S. Environmental Protection Agency, personal communication, 1986.

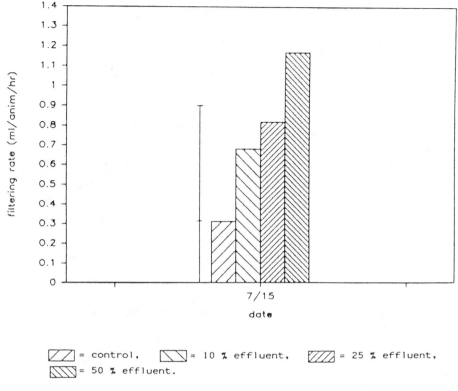

FIG. 2c—*Secondary ecosystem functional bioassay (SEFB) for Monroe WWTP effluents using* Cyclops. *The range in controls is indicated by the vertical bars.*

2. The authors were able to demonstrate the presence of toxic (inhibitory) substances at two locations, including the Monroe WWTP, from a large river system, the River Raisin, using control waters from Lake Erie. The effluents that stimulated the filtration rate came from another industrial source.

3. The inhibition of filtering was due to zinc and copper. In a surprise finding, residual chlorine stimulated the filtering rate.

4. This bioassay utilizing zooplankton filtering rates must be run in the laboratory, is very time-consuming, and involves inherent problems because of the necessity of transporting control water (including algal foods) and animals and the additional problems caused by storing effluents.

5. The SEFB has relatively low variability when directly compared with bioassays involving heterotrophs, phytoplankton, and other zooplankton. We believe that it is an acute test indicative of chronic toxicity, as reproductive effects are logically coupled to ingestive capacity.

Acknowledgments

The staff of the U.S. Environmental Protection Agency Large Lakes Laboratory in Grosse Ile, Michigan, who coordinated this study, and members of the Cranbrook Institute of Science, who collected water samples and performed chemical analyses, were especially helpful. The study was supported by U.S. EPA agreement CR 810775. The project officer, William Richard-

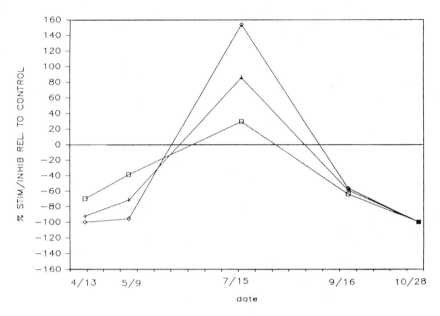

FIG. 3—*Reversal of inhibition with warm, summer weather (July) for the relative filtering rates of* Diaptomus.

TABLE 2—*Comparison of the variability of responses for controls of the four ecosystem indexes.*

Index	Date	Mean Rate (SE)	SE/mean, %	Mean Range	n
Filtration rate for *Diaptomus*,	7/83	0.42 (0.07)	16.7	0.14 to 0.70	8
mL/animal/h	9/83	0.51 (0.11)	21.6	0.29 to 1.02	6
	10/83	0.25 (0.03)	12.0	0.16 to 0.38	7
	4/84	0.39 (0.03)	7.7	0.25 to 0.56	12
	5/84	0.45 (0.02)	4.4	0.29 to 0.60	12
Average			12.5		
Net reproductive rate, R_0, for *Ceriodaphnia*, animals/female/ life table	all dates	24.0 (4.67)	19.5	13.0 to 54.0	8
Gross primary productivity of	7/83	90.9 (28.30)	31.1	40.3 to 228.2	6
Lake Erie community,	9/83	81.3 (2.78)	3.4	73.0 to 85.0	4
mg carbon/m³/h	10/83	58.5 (20.54)	35.1	30.6 to 98.6	3
	5/84	33.9 (20.05)	59.1	13.8 to 74.0	3
	6/84	52.2 (27.80)	53.3	24.4 to 80.0	2
	7/84	49.6 (17.49)	35.3	7.8 to 79.0	4
	8/84	41.5 (37.55)	90.5	3.9 to 79.0	2
Average			44.0		
Uptake of acetate by heterotrophs	9/83	1.50 (0.26)	17.3	1.14 to 2.00	3
from Lake Erie, μg acetate/L/h	10/83	−1.17 (2.05)	175.2	−5.26 to 1.05	3
	4/84	0.32 (0.25)	78.1	0.07 to 0.57	2
	5/84	0.65 (0.00)	0	0.65	2
Average			67.7		

son, deserves special thanks. Tim Anderson helped with statistics and graphs, and Terri Alston with manuscript preparation.

References

[1] Evans, M. and McNaught, D. C., "The Effects of Toxic Substances on the Characteristics of Zooplankton Populations: A Great Lakes Perspective," *Toxic Contaminants and Ecosystem Health: A Great Lakes Focus*, M. Evans, J. Gannon, D. McNaught, and C. Fetterolf, Eds., Wiley, New York, 1988, pp. 53–76.

[2] Mirza, M., "An Ecological Study on the Nature of Pollution in Tonawanda and Ellicott Creeks of the Niagara River Basin and the Effects of Various Chemical Variables on the Feeding and Reproductive Rates of *Daphnia magna*," State University of New York, Buffalo, NY, 1968.

[3] Cooley, J. M., *Journal of the Fisheries Research Board of Canada*, Vol. 34, 1977, pp. 863–868.

[4] McNaught, D. C., Griesmer, D., Buzzard, M., and Kennedy, M., "Inhibition of Productivity by PCBs and Natural Products in Saginaw Bay, Lake Huron," U.S. Environmental Protection Agency, Washington, DC, 1980.

[5] McNaught, D. C., Crane, J. L., and Meadows, C., "Effects of Toxaphene on Functional Responses of Zooplankton in Lake Superior," U.S. Environmental Protection Agency, Washington, DC, 1984.

[6] Gliwicz, M. Z. and Sieniowska, A., *Limnology and Oceanography*, Vol. 31, 1986, pp. 1132–1138.

[7] McNaught, D. C., *Journal of Great Lakes Research*, Vol. 8, 1982, pp. 360–366.

[8] Richman, S., Bohon, S. A., and Robbins, S. E. in *Evolution and Ecology of Zooplankton Communities*, American Society of Limnology and Oceanography Symposium No. 3, W. C. Kerfoot, Ed., University Press of New England, Hanover, MA, 1980, pp. 219–233.

[9] Gauld, D. T., *Journal of the Marine Biological Association of the United Kingdom*, Vol. 29, 1951, pp. 295–706.

[10] Cairns, J., *Water Research*, Vol. 15, 1981, pp. 941–942.

[11] DiToro, D. M., Connolly, J. P., Winfield, R. P., Kharkar, S. M., Wolf, W. A., Pederson, C. J., Blasland, J., and Econom, J. A., "Field Validation of Toxic Substances Model of Fate and Ecosystem Effects for Monroe Harbor," U.S. Environmental Protection Agency, Washington, DC, 1986.

[12] Mount, D. I. and Norberg, T. J., *Environmental Toxicology and Chemistry*, Vol. 3, 1984, pp. 425–434.

[13] McNaught, D. C. and Mount, D. I., *Aquatic Toxicology and Hazard Assessment: Eighth Symposium*, *ASTM STP 891*, R. C. Bahner and D. J. Hansen, Eds., American Society for Testing and Materials, Philadelphia, 1985, pp. 375–381.

S. Ian Hartwell,[1] Don Cherry,[2] and John Cairns, Jr.[3]

Fish Behavioral Assessment of Pollutants

REFERENCE: Hartwell, S. I., Cherry, D., and Cairns, J., Jr., "**Fish Behavioral Assessment of Pollutants**," *Functional Testing of Aquatic Biota for Estimating Hazards of Chemicals, ASTM STP 988,* J. Cairns, Jr., and J. R. Pratt, Eds., American Society for Testing and Materials, Philadelphia, 1988, pp. 138–165.

ABSTRACT: Avoidance of a blend of four metals (relative ratios: 1.00 copper, 0.54 chromium, 1.85 arsenic, and 0.38 selenium) by schools of fathead minnows *(Pimephales promelas)* was determined in a steep-gradient, laminar-flow laboratory chamber, an artificial stream supplied with raw river water, and a natural stream. Laboratory avoidance responses were determined seasonally during twelve months for unexposed (control) fish and for two groups of metals-acclimated fish. Field avoidance responses were determined for control fish in the artificial stream in spring and summer and in the natural stream in summer. Field avoidance responses of metals-acclimated fish were determined in the summer in the artificial stream and in the natural stream. The laboratory control fish avoided the metals blend at a concentration of 29 μg/L. The field control fish avoided 71.1 and 34.3 μg/L of total metals in the artificial stream in spring and summer, respectively, and 73.5 μg/L in the natural stream. Laboratory fish acclimated to a total metals concentration of 49 μg/L for three months did not respond to metals levels up to 245 μg/L. Laboratory fish exposed to 98 μg/L of total metals preferred elevated concentrations of 294 μg/L after three months of exposure, mildly avoided 490 μg/L after six months, and were not responsive to concentrations approaching 980 μg/L after nine months of exposure. Field-acclimated fish did not respond to metal blends as high as 1470 and 2940 μg/L in the artificial and natural streams, respectively. Water hardness, turbidity, and physical setting are implicated as possible causative factors in the differences in field avoidance levels among the control fish. The 96-h median lethal concentration (LC_{50}) of the metals blend for acclimated fish was 1.25 and 1.41 times higher than that for control fish for the laboratory and field groups, respectively.

KEY WORDS: fish avoidance, field validation, metals acclimation, synergism, toxicity, hazard evaluation

Responses of fish to various chemical and physical parameters in the aquatic environment are becoming increasingly important as a consequence of enhanced public concern for fish stocks as a recreational and economic resource, for water quality in general, and also as a consequence of legal regulation. The Toxic Substances Control Act of 1976 stipulated that new chemical substances, prior to their manufacture and distribution, must be evaluated for their effects on human health and the environment, including the toxicological and environmental fate of the chemical, among other effects [1]. Acute toxicity bioassays are used for screening chemical potency; however, predictive results cannot be obtained and the relevance of such data to aquatic ecosystems in the real world is not adequately resolved by this method; therefore, environmental

[1]Research associate, University of Maryland, Horn Point Environmental Laboratories, Cambridge, MD 21613.

[2]Professor, Department of Biology, Virginia Polytechnic Institute and State University, Blacksburg, VA 24061.

[3]Director, University Center for Environmental Studies, Virginia Polytechnic Institute and State University, Blacksburg, VA 24061.

concern levels are estimated by "assessment factors" [2]. Acute lethality bioassays cannot address questions of sublethal effects; likewise, chronic exposure to some pollutants may result in cumulative or delayed effects. Simple laboratory acute toxicity tests cannot adequately predict synergistic effects of various materials or easily incorporate variation in the background water quality into the tests. In short, acute lethality tests are useful as a yardstick for comparing relative toxicity, but their results cannot be reliably extrapolated to predict environmental consequences.

Both sublethal and chronic exposure experiments have addressed a number of questions pertinent to environmental sensitivity to pollutants. It has been demonstrated that long-term exposure may have severe impacts on growth and reproductive success [3-7]. Chronic exposure can also result in direct toxicity or reduced resistance to disease and predators [8-11]. In those studies for which comparative data are available, chronic exposure effects often occur at concentrations well below acute levels. The implication of these observations is that the environment's ability to resist chemical insult is drastically lower than the levels suggested by the simple survival of organisms. The integrity of a robust ecosystem is crucial to its ability to assimilate stressful inputs [12].

Direct toxic action of pollutants may occur at release points, but after the material in question has spread out and become diluted, vastly larger and more diverse habitats and their residents may be exposed to long-term sublethal exposure. Materials that may affect the normal functioning of neurosensory systems are of interest, since they may affect how fish move through the environment and respond to the normal range of cues which direct them toward food, shelter, spawning grounds, and so forth. The proper response to those cues and the ability to avoid harmful situations are of paramount importance to the survival of the population. The ability to predict what levels of a pollutant will cause harmful side effects in the organisms in a receiving system is limited by the ability of laboratory-derived results to be extrapolated to the real environment.

While numerous researchers have attempted to contrast laboratory results with field distributions of aquatic animals exposed to pollution, no one has attempted to recreate their laboratory experiments in a natural setting. This endeavor would generate information on the validity of laboratory studies and lead to information concerning how and why organisms respond in the way they do to effluents and normal environmental cues.

Clear demonstration of laboratory avoidance at levels below those causing harmful effects may be meaningless from an environmental standpoint if the organisms fail to avoid those concentrations in the wild. Conversely, if organisms avoid field concentrations below harmful levels, the result is a de facto loss of an otherwise suitable habitat. Highly variable conditions in the environment over seasonal time spans and the variable nature of industrial effluents may render laboratory results questionable as indicators of environmental harm without concurrent field validation. Well-defined avoidance response bioassays need to be developed and verified with field tests since aquatic organisms respond to the environmental as a composite, with some stimuli being more important than others. Because the "environment" cannot be completely duplicated in the laboratory, correlations between important stimuli in the laboratory and the field must be derived for purposes of hazard evaluation and regulatory control. Improved knowledge of the actual environmental effects of effluents, through generation of scientifically acceptable data, will lead to improved understanding and therefore more optimal use of the environmental system.

Objectives

The objectives of this study were (1) to determine avoidance concentrations of a metals combination in a steep-gradient laboratory avoidance chamber for each of three groups of laboratory-maintained fish (the groups were exposed constantly to either a control, low, or high metals concentration); (2) to determine avoidance concentrations of the metals combination in an artificial stream supplied with raw river water, for each of two groups of fish maintained in river water and exposed constantly to either a control or high metals concentration; and (3) to determine avoidance concentrations of the metals combination in a natural stream for each of the same two groups of fish; and (4) to determine the acute toxicity of the metals combinations to exposed and unexposed fish from the laboratory and river water holding facilities.

Materials and Methods

Laboratory Exposure

The fish were maintained in 400-L fiberglass tanks (Fig. 1), previously described by Hartwell et al. [13]. The photoperiod and temperature were altered to mimic seasonal cycles. Oxygen saturation was measured at irregular intervals and was always above 90%. The temperature varied between 5 and 25°C depending upon the season. The test temperature did not vary more than ±2°C between exposure groups. Two hundred fifty fish were introduced into the holding tanks in September 1982. In January 1983, two of the tanks began receiving doses of a blend of three heavy metals (copper, cadmium, and chromium) and two metalloids (arsenic and selenium). The relative concentrations were derived from the observed concentrations found in a fly-ash slurry entering a settling basin at a small coal-fired power plant in Virginia [14]. The average slurry concentrations were as follows: copper, 1.3 mg/L; chromium, 0.7 mg/L; cadmium, 0.03 mg/L; arsenic, 2.4 mg/L; and selenium, 0.5 mg/L. The initial exposure levels were 0 (control), 5% (247 μg/L metals), and 10% (493 μg/L metals) of the fly-ash slurry content. Exposure levels were maintained by renewing 20% of the water weekly and adding the appropriate amount of metals. Water samples for metals analysis were taken weekly, preserved at pH 2 with redistilled nitric acid (HNO_3). During the first season of dosing and periodically thereafter, water samples were also taken 24 h after water renewal to ensure that the metals levels did not fluctuate after dosing. The percentage of dissolved metals was greater than 90% at all times in the holding tanks and in the avoidance chamber. In addition, several water-quality characteristics were routinely monitored (Table 1).

After six weeks, both treated groups experienced mortalities, and significant numbers of fish were displaying equilibrium loss, abnormal color, and no response to food. The exposure schedule was reduced to 0 (control), 1% (49 μg/L of metals), and 2% (98 μg/L of metals) of the fly-ash slurry content, and cadmium was removed altogether because it was below reliable detection limits (Table 1).

The metals were analyzed by the furnace technique on a Perkin-Elmer Model 703 atomic absorption spectrophotometer equipped with a deuterium-arc background lamp for arsenic and selenium. The analytical methods and quality control procedures followed U.S. Environmental Protection Agency (EPA) methods [15,16].

Field Exposure

Two groups of fathead minnows were kept in separate 400-L fiberglass tanks as just described. The tanks were located in a laboratory trailer on the premises of the Appalachian Power Company's steam-generating plant in Glen Lyn, Virginia. The water supply was raw New River water from the intake well in the power plant screen house located upstream from the plant effluents. Fish were introduced into the holding tanks in February 1984. After two months of acclimation, one group was dosed with a blend of the four metals equal to the high-dose concentration in the laboratory exposure portion of the research. The other group served as controls.

The water sampling and metals exposures were performed by the same procedures used in the laboratory exposure tests. The holding water samples and subsequent avoidance test water sam-

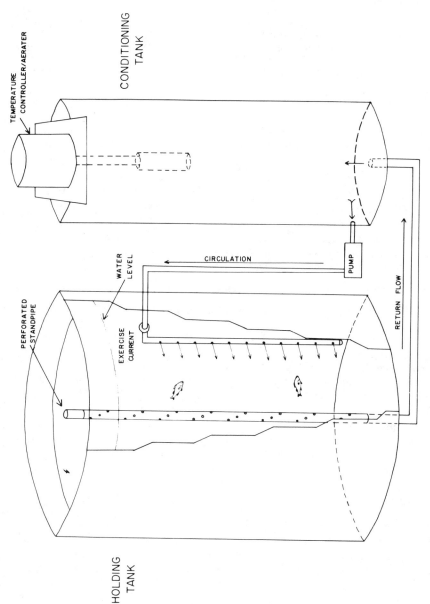

FIG. 1—*Circular holding tanks used for acclimating fish to heavy metal blends for avoidance experiments in the laboratory at Virginia Polytechnic Institute, using dechlorinated tap water, and at Glen Lyn, Virginia, using raw New River water.*

TABLE 1—*Water quality parameters of the holding water for fathead minnows during the 15-month holding period.*[a]

Parameter	Control	Low Dose	High Dose
Nonmetals, mg/L of the ion[b]			
NO_3	4.19 (2.89)	2.26 (0.85)	6.13 (4.38)
NH_3	0.076 (0.032)	0.064 (0.036)	0.152 (0.133)
PO_4	0.310 (0.183)	0.177 (0.119)	0.466 (0.039)
SO_4	11.16 (3.77)	6.85 (3.34)	12.58 (4.19)
Cl	7.92 (1.65)	8.70 (2.73)	7.49 (2.33)
Hardness, mg/L $CaCO_3$	89.0 (32.5)	85.6 (26.0)	79.0 (27.7)
Alkalinity, mg/L $CaCO_3$	32.0 (7.69)	28.8 (5.34)	36.7 (4.53)
pH	7.5 (0.3)	7.3 (0.2)	7.6 (0.2)
Metals, mg/L			
Cu^c	0.011 (0.006)	0.012 (0.003)	0.009 (0.004)
Cr^c	BD[d]	BD	BD
Cd^c	BD	BD	BD
As^c	BD	BD	BD
Se^c	BD	BD	BD
Cu^e	0.011 (0.006)	0.034 (0.006)	0.031 (0.009)
Cr^e	BD	0.004 (0.003)	0.006 (0.002)
As^e	BD	0.018 (0.006)	0.025 (0.005)
Se^e	BD	0.011 (0.005)	0.008 (0.006)

[a]The numbers in parentheses are the standard deviations.
[b]For nonmetals, $n = 20$; for low-dose nonmetals, $n = 10$.
[c]Before dosing began, $n = 5$.
[d]BD = below the detection limit (copper, 0.002; chromium, 0.002; arsenic, 0.005; and selenium, 0.005 mg/L).
[e]After concentration adjustment, for the control and high dose, $n = 28$; for the low dose, $n = 11$.

ples required wet digestion with redistilled HNO_3 according to EPA methods [15] to attain 100% recovery in the analyses. In addition, the water samples were filtered (0.45 μm) for determination of the dissolved metals concentration, which was always above 90% in the holding tanks and during testing. Several water quality characteristics were routinely monitored (Table 2). The photoperiod was natural daylight and the temperature was controlled to mimic seasonal cycles. The fish were fed twice a day with automatic feeders. The temperature varied between 5 and 24°C, depending on the season. Oxygen saturation was measured at irregular intervals and was always above 90%.

Avoidance Testing

Laboratory avoidance of the metals by fish was determined in a steep-gradient, laminar-flow chamber (Fig. 2), previously described by Hartwell et al. [13]. The chamber was constructed of clear acrylic plates. One end was 50 by 50 cm in area and 50 cm deep at the deepest point. It was bisected by a clear divider. Each side was supplied with dechlorinated tap water through a perforated "T" spreader. The water in the deep end was thoroughly mixed before flowing through screens into a 50 by 50-cm fish holding area. Dye testing demonstrated that the water flow was virtually laminar through this area. Test solutions injected into one of the mixing sections flowed across half of the observation area, creating a boundary layer down the midline. The test fish then choose where to reside.

The chamber was enclosed, and a video camera was located above the observation area. A video monitor located away from the chamber was used for data collection. Control fish were tested in 1983 during the spring (after 6 months of holding), summer (after 9 months of hold-

TABLE 2—*Water quality parameters of the holding water for fathead minnows during the seven-month holding period in New River water.*[a]

Parameter	Control[b]	Exposed
Nonmetals, mg/L of the ion[c]		
NO$_3$	41.2 (18.2)	50.9 (14.0)
PO$_4$	1.64 (1.23)	1.87 (1.32)
SO$_4$	13.9 (4.3)	15.2 (4.9)
Cl	4.6 (1.4)	4.9 (1.0)
Hardness, mg/L CaCO$_3$	76.7 (8.3)	84.3 (8.0)
Alkalinity, mg/L CaCO$_3$	26.5 (9.8)	19.9 (7.6)
pH	7.3 (0.4)	7.1 (0.4)
Turbidity, Jackson turbidity unit	1.4 (0.5)	1.3 (0.3)
TOC, mg/L	6.1 (1.5)	5.2 (0.7)
Metals, mg/L[d]		
Cu	0.012 (0.004)	0.040 (0.013)
Cr	0.003 (0.002)	0.011 (0.002)
As	BD[e]	0.053 (0.013)
Se	BD[e]	0.008 (0.003)

[a]The numbers in parentheses are the standard deviations.
[b]For control fish, $n = 21$.
[c]For nonmetals, $n = 21$.
[d]For metals, $n = 16$.
[e]BD = below the detection limit (copper, 0.002; chromium, 0.002; arsenic, 0.005; and selenium, 0.005 mg/L).

ing), and fall (after 12 months of holding) of 1983 and the winter (after 15 months of holding) of 1984. The low-exposure (1%) tests were performed in the summer of 1983, 3 months after the onset of the change in the exposure. The high-exposure (2%) tests were performed in the summer and fall of 1983 and the winter of 1984 after 3, 6, and 9 months of metals exposure at the new levels. Because of time constraints and the necessity for equal acclimation periods for the control and exposed fish, the survivors in the exposed groups were not replaced after the dosage reductions.

The tests were conducted by placing a school of five fish in the test chamber and allowing them to calm down for 30 to 60 min and exhibit random exploratory activity. A 10-min control period followed with no metals solution input. The location of each fish was recorded at 30 s intervals. After the control period, a metals-blend solution, made up of the same relative concentrations as the long-term exposure blend, was injected into one side of the chamber by a peristaltic pump. After 5 min of dosing, another 10-min test period was conducted, and observations were made of the fish locations every 30 s. This procedure was followed so that the test concentration increased in a stepwise progression to levels that were roughly linear on a log scale.

Control fish were tested at nominal concentration levels from 0 (control) to 2.2% (108 μg/L metals) of the concentration of the fly-ash basin. Low-exposure fish were tested at concentration levels from 0 (control) to 6.6% (323 μg/L metals). High-exposure fish were tested at nominal concentration levels from 0 (control) to 4.6% (225 μg/L metals), 0 to 16% (748 μg/L metals), and 0 to 23.0% (1127 μg/L) in summer, fall, and winter, respectively. The target concentrations varied seasonally in the high-dose trials because the tests were run until a response was definite or to the limit of the dosing setup capabilities.

Water samples were drawn from the test area at each concentration by means of sample ports located in the floor of the test area. In tests in which fish from the long-term exposure groups were used, a background concentration of metals equivalent to their long-term exposure level

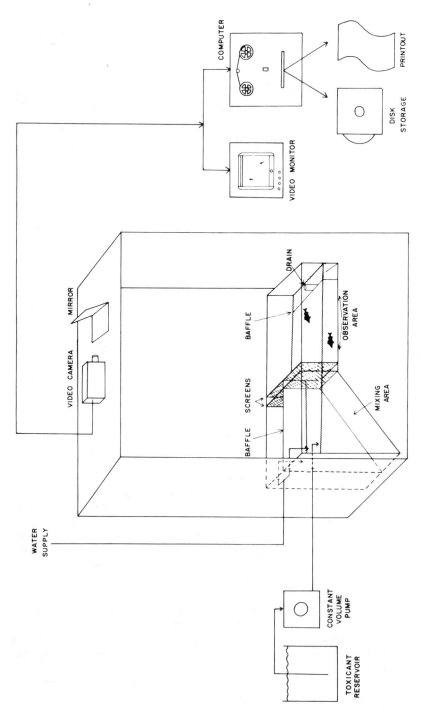

FIG. 2—*Fish avoidance chamber with a video monitor/computer interfaced system.*

was injected into the avoidance chamber water supply so that the animals did not experience a novel situation relative to the metals concentration. After each set of tests, the fish were discarded, so novice fish were used in each replicate in each season. All tests were replicated at least five times except the summer controls tests. Because of an accident in an adjacent laboratory, all the low-dose (1%-exposure) fish perished, and the control fish holding tank was damaged. The tests were suspended, until the fall season, before the summer control group tests were completed.

Field Avoidance Testing

Avoidance responses were determined in a modified artificial stream supplied with water from the New River, Virginia, and in a natural stream setting in Adair Run, a small second-order tributary to the New River. Control fish were tested beginning in April in the artificial stream and during July and August in the artificial stream and in Adair Run. Metals-acclimated fish were tested in the artificial stream in July (after three months of metals acclimation) and in Adair Run beginning in September (after five months of metals acclimation). The Adair Run tests were delayed until September because of low water in the latter half of August.

The artificial stream was constructed of a galvanized hatchery trough (20 cm deep by 39 cm wide by 4.6 m long). The inside was coated with white epoxy paint. New River water was pumped from the screen-house intake well into one end of the trough and flowed through a series of baffles and screens to smooth the flow profile (Fig. 3). A concentrated solution of metals was pumped into the mixing zone upstream of the screens through a perforated "T" manifold to dose one half of the trough. Water flowed through the observation area in nearly laminar fashion with a boundary down the middle. Dye testing demonstrated excellent separa-

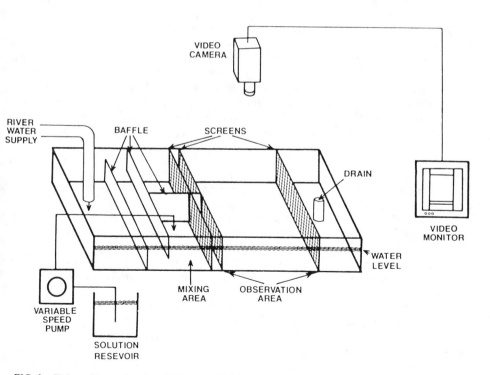

FIG. 3—*Fish avoidance chamber within an artificial stream for testing the responses of fathead minnows to metal-blend solutions in raw New River water. The canopy is not illustrated.*

tion of the two parcels of water all the way through the observation area. The observation area measured 39 by 39 cm and the depth was maintained at 5 cm. The flow rate was regulated at 20 L/min. A video camera was mounted on the canopy frame directly above the observation area and fed into a video monitor located in the laboratory trailer for remote observation. Sample ports located in the floor of each side of the observation area led to stopcocks outside the canopy, which allowed the water to be sampled during testing without disturbing the fish. During tests for the fish with long-term metals acclimation, a background concentration of the metals blend equal to their acclimation level was injected into the water supply for the artificial stream, as in the laboratory procedures.

Control fish were tested at nominal concentration levels from 0 (control) to 2.2% (108 μg/L metals). The fish with long-term metals acclimation were tested at nominal concentrations from 0 (control) to 20% (980 μg/L metals) of the fly-ash basin influent concentration, above background (2% exposure) levels.

In-Stream Avoidance Testing

The avoidance trials in the natural stream setting were conducted in Adair Run. It received the effluent from the power plant's fly-ash pond ~75 m above its confluence with the New River (Fig. 4). The avoidance tests were conducted in shallow pools ~1 km above the New River confluence. The stream basin above the test area has no known urban or agricultural runoff point sources. In the test area, the stream bottom was flat slab rock with patchy areas of gravel and cobble. The avoidance trials were conducted in a nylon mesh enclosure placed directly in the stream (Fig. 5). The enclosure was constructed of a polyvinyl chloride frame 50 cm wide by 44 cm long at the base with 45° sloping sides. A nylon net (3.8-mm mesh) was stretched over the frame and secured at the four upright corners. The enclosure was surrounded on three sides by a black plastic canopy by was open on the upstream side. An observation slit was cut in the downstream canopy wall so an observer could stand downstream of the enclosure and watch the fish from above without disturbing them. A battery-powered, variable-speed pump injected the metals blend into a manifold located upstream of the enclosure which was positioned so that one half of the enclosure was dosed. No attempt was made to control the flow rate, turbulence, or any aspect of stream flow through the enclosure; however, meticulous dye testing prior to each replicate was necessary to position the enclosure so that the stream flowed straight through the enclosure and the metals solution did not meander to the other side. At times, this proved quite difficult, and the exact test location varied by several metres from day to day so that suitable flow patterns and flow rates could be located. Dye tests were also conducted following each replicate because the flow rates and patterns could vary over a matter of hours if it had rained during the previous 24 h. The flow volume was measured prior to each replicate and varied between 26 and 120 L/min (mean flow rate = 58 L/min). Each side of the enclosure had a sampling tube affixed in the center. Water samples were taken directly from the two sides of the test area by drawing water through a silicone tube with a 50-cm³ syringe. In tests using fish with long-term metals acclimation, a background concentration of the metals blend equal to their exposure level was introduced into the stream above the test area with a Mariotte bottle. Again, placement was guided by dye tests, and the flow rate was based on the measured stream flow upstream of the enclosure. Experiments were conducted in the stream enclosure and the artificial stream using the same procedures as in the laboratory.

Control fish were tested at nominal concentration levels from 0 (control) to 2.2% (108 μg/L metals), and the metals-acclimated fish were tested at nominal concentrations from 0 (control) to 20.0% (980 μg/L metals) of the fly-ash basin influent concentration, above background (2% exposure) levels.

The fish usually showed a preference for one side or the other of the test area during the control periods, and in those cases, the dosing manifold was placed on that side at the beginning of the test. When an avoidance response was evident, the position of the manifold was switched

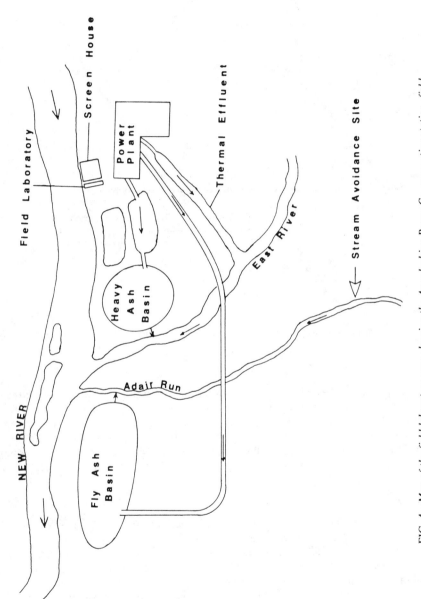

FIG. 4—*Map of the field laboratory area showing the Appalachian Power Co. generating station, field laboratory, screen house, fly ash basin, and Adair Run field site.*

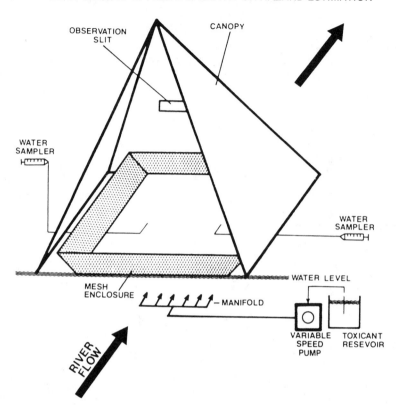

FIG. 5—*Portable avoidance chamber for testing responses of fathead minnows to metal-blend solutions in a natural stream setting.*

to the opposite side at the end of the test series for another 10-min period to confirm the response. Water samples were drawn from the test areas at each test concentration by means of the sampling tubes. In addition to metals analysis, water samples were drawn for analysis of nitrate (NO_3), sulfate (SO_4), phosphate (PO_4), chloride (Cl), hardness, alkalinity, pH, turbidity, and total organic carbon (TOC). Following each test, the fish were discarded, so each replicate in each season used novice fish. All the tests were replicated at least eight times.

Toxicity Testing

Acute lethality bioassays were conducted on four groups of fish: (1) laboratory controls, (2) high-metals-acclimated fish, (3) field controls, and (4) field-metals-acclimated fish. These tests were conducted after all avoidance experiments were concluded. The bioassay test fish were the fish remaining after all the avoidance-tested fish had been removed. In the case of the laboratory high-metals-acclimated fish, a group different from those used in the avoidance trials was used. The second group of fish was originally slated to be used to repeat the three-month exposure to the high metals blend, without an initial cadmium spike. This experiment was abandoned because of the lack of arsenic supplies, which lasted six weeks at the end of the three-month exposure. However, the second group of fish was maintained a full ten months (24 May 1984–25 March 1985) under exposure for purposes of the bioassay.

The 96-h flow-through bioassays were conducted in a solenoid-activated serial diluter, originally described by Hendricks et al. [17]. The diluter had adjustable flow rates and solution concentrations according to the target test concentrations. The acute toxicity testing followed EPA recommended methods [18]. The reported values for each replicate test are averages of daily measurements.

In addition to the four-metal bioassays, a bioassay was conducted using all five of the elements in the original blend. Cadmium was added to the stock solutions as cadmium chloride $(CdCl_2 \cdot 2.5H_2O)$.

Statistical Evaluation

A detailed discussion of the development of the statistical methods may be found in Hartwell et al. [19]. The total number of fish counts on the dosed side was summed for each 10-min period. With five fish and 20 counting periods, the total possible was 100 counts. Quadratic regressions of the proportion of fish on the dosed side versus the test concentration and an arcsine transformation of the proportion versus the test concentration were calculated to test for deviations from normality. The proportion of fish on the dose side was used as the dependent variable in least squares quadratic regressions against the test concentration using replicates designated as dummy variables having the values 1 or 0 to test for parallelism and coincidence of replicate lines. Parallel lines indicated that the avoidance response was the same from replicate to replicate. Coincident lines indicated that the avoidance response occurred at the same concentration in each replicate. The laboratory control regression lines were parallel and coincident, and they were pooled within each season and tested for differences between seasons by analysis of variance. Similarly, the control data from the field trials were pooled and tested for differences between seasons and between test apparatus, as well as against the laboratory control data. For field data, the measured water quality parameters were used as covariates in the analysis of variance comparisons between test systems (artificial stream versus Adair Run). The bioassay data were analyzed using standard probit analysis. All statistical tests were performed with the SAS [20] computer program.

Results

Laboratory Avoidance

The experiments clearly demonstrate that fathead minnows are sensitive to metal blends. Laboratory control fish strongly avoided low concentrations of the metal blend (Fig. 6). Regression and correlation coefficients from regressions of residence versus the test concentration are shown in Table 3. Analysis of variance between seasons indicated that no significant differences existed between responses over the 12-month period ($F = 0.39$; degrees of freedom (df) = 4,83). The pooled, mean concentration at which the residence in the dosed side was half that of the control period was 0.63% (31 $\mu g/L$ metals) of the fly-ash basin influent concentration (standard deviation = 0.23) (Table 4). The low-exposure (1%) group did not respond to levels of the blend elevated five times higher than acclimation levels (Fig. 7, Table 4). The regression coefficient of the test concentration versus residence is not significantly different from zero (Table 3). The variability is very high, as is indicated by an r^2 value below 10%.

The high-exposure (2%) fish had variable responses, depending on the length of the metals acclimation period (Fig. 8). After three months' exposure to the reduced concentration, the fish were attracted to levels of metals three times higher than the acclimation level (Table 4). The quadratic regression F value is highly significant, but the correlation is not as high as control test correlations. After six months' acclimation the overall response was reversed (Fig. 8), with the fish displaying a mild avoidance response at high concentration levels. The quadratic regression F value is significant (Table 3) at the $\alpha = 0.01$ level, and the correlation is lower than for the previous season. After nine months' acclimation, the fish were indifferent to low or high concentrations (Fig. 8). The regression F value is significant at $\alpha = 0.05$, and correlation is

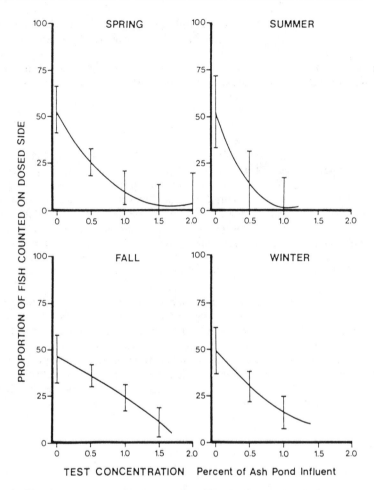

FIG. 6—*Avoidance responses of fathead minnows to a blend of four metals after 6 (spring), 9 (summer), 9 (fall), and 15 (winter) months of laboratory acclimation in uncontaminated water. The vertical bars show the 95% confidence limits of the quadratic regression.*

lower than for both previous seasons (Table 3). The replicate lines are statistically parallel but not always coincident, which indicates that the responses were not consistent from group to group.

Field Avoidance

The holding water quality characteristics and metals concentrations during the experimental period are shown in Table 2. Most of the parameters were stable throughout the entire period. The nitrate, PO_4, and, to a lesser extent, SO_4 and chlorine increased gradually from February to late May and early June, and then declined slightly. These values compare well with river characteristics during the test periods (Tables 5a and 5b). Nitrate, PO_4 and TOC were elevated in the holding tanks because of the presence of 250 fish and daily feeding. Likewise, the alkalinity and pH were slightly depressed because of metabolic waste products from the fish. Copper was

TABLE 3—*Regression coefficients for multiple quadratic regressions of fathead minnow residence times on the exposed side of avoidance apparatus versus the measured concentration.*[a]

Exposure	Season	Concentration	Concentration Squared	r^2
			Regression Coefficient	
Control	spring	-60.9^b	18.2^b	55.3
	summer	-99.3^b	47.6^b	73.3
	fall	-17.2^b	-3.7^b	54.2
	winter	-42.9^b	10.3^b	47.9
Low dose	summer	-13.7	2.0	9.5
High dose	summer	-0.8^b	1.8^b	36.1
	fall	-5.7^b	0.3^b	28.6
	winter	0.6	0.0	17.6

[a]The regression coefficients for linear and quadratic concentration terms and the overall correlation coefficient, r^2, are shown.
[b]Significant at the $P < 0.01$ level.

slightly elevated in the holding tanks, probably as a result of leaching from the heat-exchange coils on the water coolers.

The long-term exposure group was acclimated to an average blend concentration equal to 1.8% (88 μg/L metals) of the fly-ash basin influent concentration (target concentration = 2.0%), above background levels.

As a result of elevated background concentrations of chromium and copper from an unknown source and of a transient nature, three of eight spring control replicates and four of nine summer control replicates in the artificial stream were discarded. Also, because of changed flow patterns and eddy formation in Adair Run, one of the eight control trials was discarded. The crude measurement of the stream flow in both systems consistently overestimated the dilution volume; so, measured test concentrations were higher than the target levels. They were not high enough to compromise the experimental design, however.

The avoidance experiments clearly demonstrate that unexposed fish avoid metals solutions in the field (Table 4). Control fish avoided the metals blend at relatively low levels in both seasons and both test systems (Figs. 9 and 10). The concentration for which the residence was one half of the control period was 1.45% (71 μg/L metals; 95% confidence interval (CI) = 0.40%) and 0.70% (34 μg/L metals; 95% CI = 0.15%) for the spring and summer artificial stream trials, respectively, and 1.50% (74 μg/L metals; 95% CI = 0.50%) in the Adair Run trials. The mean control period residence prior to introduction of the metals varied between 68 and 70% over all three locations. Individual replicate lines within each test system were parallel but not always coincident (Table 6). The pooled slope of the summer artificial stream trials is significantly steeper than the slopes for the spring artificial stream and the summer Adair Run trials ($F = 3.22$; df = 2,58; and $F = 3.91$; df = 2,62, respectively). The spring artificial-stream and summer Adair Run slopes were not significantly different ($F = 0.66$; df = 2,68).

Two-way analyses of variance (ANOVAs) of avoidance with the test system and concentration as independent variables were calculated with and without the inclusion of water quality parameters as covariates. The probability values for significance of the test system are shown in Table 7. Only the addition of hardness explains a significant portion of the difference in avoidance responses between all three systems.

The acclimated fish did not respond to the metals blend in either test system (Figs. 9 and 10).

TABLE 4—Avoidance test levels expressed as a percentage of the fly-ash basin influent concentration and the element concentrations at those levels.[a]

Exposure History	Avoidance Test	% of Influent Concentration	Concentration, μg/L			
			Copper	Chromium	Arsenic	Selenium
Laboratory control	laboratory	0.63[b]	8	4	15	3
Laboratory exposed						
Low dose,						
3 months	laboratory	5.00 NR[c]	65	35	120	25
High dose						
3 months	laboratory	5.00[d]	65	35	120	25
6 months	laboratory	14.00 NR	182	98	336	70
9 months	laboratory	20.00 NR	260	140	480	100
Field control						
Spring	artificial stream	1.45	19	10	35	7
Summer	artificial stream	0.70	9	5	17	4
	Adair Run	1.50	20	11	36	8
Field exposed,						
summer	artificial stream	30.00 NR	390	210	720	150
	Adair Run	60.00 NR	780	420	1440	300

[a]Where no response was evident, the highest levels tested are shown.
[b]Pooled data.
[c]NR = No response.
[d]Showing 50% attraction.

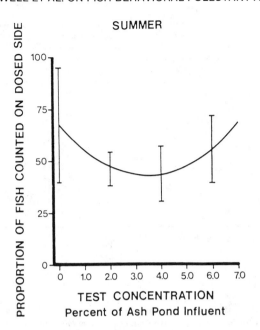

FIG. 7—*Avoidance responses of fathead minnows to a blend of four metals after three months of exposure to a concentration equal to 1% of the concentration found in a fly-ash basin influent. Test concentrations are expressed as the test concentration plus background exposure. The vertical bars show the 95% confidence limits of the quadratic regression.*

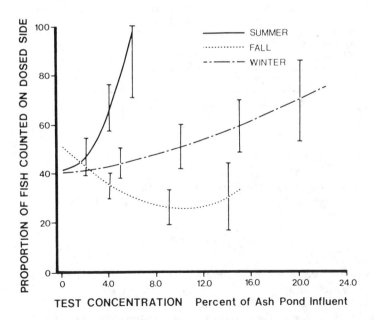

FIG. 8—*Avoidance responses of fathead minnows to a blend of four metals after three (summer), six (fall), and nine (winter) months of exposure to a concentration equal to 2% of the concentration found in a fly-ash basin influent. Test concentrations are expressed as the test concentration plus background exposure. The vertical bars show the 95% confidence limits of the quadratic regression.*

TABLE 5a—*Water quality parameters of the New River during field testing of fathead minnow avoidance of heavy metal blends.*[a]

Parameter	New River Control, Spring[b]		New River Control, Summer[b]		New River Exposed, Summer[c]	
Nonmetals, mg/L of the ion						
NO$_3$	5.4	(2.6)	6.2	(3.6)	5.6	(2.4)
PO$_4$	0.04	(0.08)	0.00	(0.00)	0.00	(0.00)
SO$_4$	10.4	(1.9)	17.4	(9.6)	13.6	(4.2)
Cl	2.7	(0.7)	4.4	(2.2)	2.8	(0.7)
Hardness, mg/L CaCO$_3$	78.0	(4.5)	70.0	(14.1)	67.5	(12.8)
Alkalinity, mg/L CaCO$_3$	50.0	(0.6)	53.2	(1.8)	42.8	(7.6)
pH	7.8	(0.4)	7.9	(0.1)	7.7	(0.4)
Turbidity, Jackson turbidity unit	5.3	(1.5)	3.2	(1.1)	6.1	(2.4)
TOC, mg/L	1.6	(0.1)	1.9	(0.4)	2.4	(0.3)

[a]The numbers in parentheses are the standard deviations.
[b]$n = 5$.
[c]$n = 8$.

TABLE 5b—*Water quality parameters of Adair Run during field testing of fathead minnow avoidance of heavy metal blends.*[a]

Parameter	Adair Run Control, Summer[b]		Adair Run Exposed, Summer[c]	
Nonmetals, mg/L of the ion				
NO$_3$	1.7	(0.6)	5.8	(6.4)
PO$_4$	0.00	(0.00)	0.00	(0.00)
SO$_4$	18.2	(5.4)	26.0	(2.7)
Cl	5.5	(0.5)	5.2	(0.9)
Hardness, mg/L CaCO$_3$	100.3	(18.0)	141.3	(8.3)
Alkalinity, mg/L CaCO$_3$	76.3	(9.6)	117.5	(8.5)
pH	7.8	(0.1)	7.8	(0.2)
Turbidity, Jackson turbidity unit	10.2	(3.2)	2.6	(0.3)
TOC, mg/L	2.8	(0.4)	2.4	(0.3)

[a]The numbers in parentheses are the standard deviations.
[b]$n = 7$.
[c]$n = 8$.

The slopes of residence versus metals concentration are not significantly different from zero (Table 6). The correlation coefficients are extremely low and individual replicate lines are not parallel (Table 6).

Toxicity

Ninety-six-hour median lethal concentration (LC$_{50}$) values are shown in Table 8. The high-dose laboratory exposed group was 1.25 times more resistant to the metals blend than the laboratory control fish. The field exposed group was 1.41 times more resistant to the metals blend than the field control group. The two control groups were equally resistant to the metals blend, as were the two acclimated groups. The 96-h LC$_{50}$ value for the five-metal blend experiment was only 80% of the concentration of those elements in the fly-ash basin (Table 8), or approximately one half of the 96-h LC$_{50}$ of the four-metal blend.

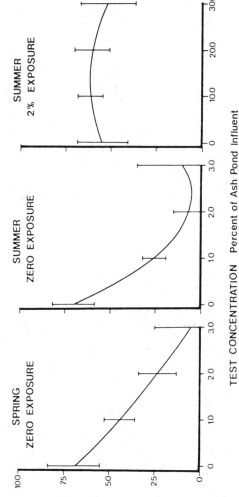

FIG. 9—Avoidance responses of fathead minnows in an artificial stream to a blend of four metals. The zero-exposure fish are control fish maintained in unaltered New River water for three (spring) and six (summer) months. The 2%-exposure fish were acclimated in New River water to a metal-blend concentration equal to 2% of the concentration found in a fly-ash basin influent. Test concentrations are expressed as the test concentration plus background exposure. The vertical bars show the 95% confidence limits of the quadratic regression.

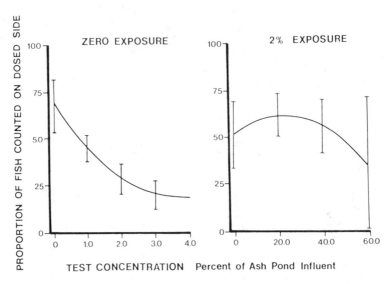

FIG. 10—*Avoidance responses of fathead minnows in Adair Run to a blend of four metals. The zero-exposure fish are control fish maintained in unaltered New River water for three months. The 2%-exposure fish were acclimated for five months in New River water to a metal-blend concentration equal to 2% of the concentration found in a fly-ash basin influent. Test concentrations are expressed as the test concentrations plus background exposure. The vertical bars show the 95% confidence limits of the quadratic regression.*

Discussion

Laboratory Avoidance

The consistency of the control fish responses in the laboratory tests over four seasons indicated that behavioral tests are an effective method for testing for effects of metals in the absence of prior exposure. Another aspect of the control fish behavior in the laboratory is the lack of threshold in the response, in spite of the low concentrations used in these tests. A decline in residence on the dosed side begins at the lowest levels and continues throughout the tests. Similar results have been reported for copper [21] and zinc [22] tested individually.

At an arbitrary cutoff point of half the control period residence, the average metals concentration was 0.63% (31 μg/L of metals) of the fly-ash basin influent concentration (Table 4), or 0.4% of the 96-h LC_{50} value (Table 8). This is equivalent to 8.1 μg/L of copper, 4.4 μg/L of chromium, 15.0 μg/L of arsenic, and 3.1 μg/L of selenium. These values compare favorably with literature values of metals tested individually. Copper has been most frequently studied in this regard. Fish avoid levels from 0.1 to 74 μg/L of copper [21,23–25]. Also, depending upon the test method and species, copper may be ignored or preferred at high concentrations, yet avoided at low concentrations [21–25]. Sprague [26] also reported that copper and zinc were synergistic in eliciting an avoidance response. Zinc is avoided by largemouth bass *(Micropterus salmoides)* at levels at least as low as 7 mg/L and by rainbow trout at 0.5 mg/L, but is not avoided by bluegill sunfish at levels between 11 and 43 mg/L [25]. However, Sprague [22] reported that rainbow trout avoided zinc concentrations as low as 5.6 μg/L. Whether this difference is due to methodology or background water characteristics cannot be determined. Rainbow trout avoid nickel at 23.9 μg/L [21] and cadmium at 52 μg/L but were attracted to mercury at 0.2 μg/L [25]. Thus, the behavioral response to the interaction of metals blends is highly

TABLE 6—F values and regression coefficients for multiple quadratic regressions of fathead minnow residence times on the exposed side of avoidance apparatus versus the measured concentration.

Test Group	Test System	F Values		Regression Coefficient		
		Parallelism[a]	Coincidence[b]	Concentration	Concentration Squared	r^2
Spring control	artificial stream	0.85	3.51[c]	−26.21[c]	1.63	45.7
Summer control	artificial stream	2.48	0.45	−75.32[d]	16.88[c]	73.5
Summer exposed	artificial stream	3.56[d]	[e]	1.03	−0.04	6.2
Summer control	Adair Run	1.95	3.54[d]	−27.30[d]	3.65	44.3
Summer exposed	Adair Run	2.60[c]	[e]	0.86	−0.02	2.8

[a]Significant F values indicate that the replicate lines are not parallel.
[b]Significant F values indicate that the replicate lines are not coincident (that is, they have different intercepts).
[c]Significant at the 0.05 level.
[d]Significant at the 0.01 level.
[e]If lines are not parallel, by definition they cannot be coincident.

TABLE 7—P *values for significance of the test systems (artificial stream in spring or summer, and Adair Run) with and without inclusion of water quality parameters as covariates in an analysis of variance model of fish residence versus metals concentration for fathead minnows.*

		Test System Comparison	
	All Locations	Artificial Stream, Spring Versus Summer	Adair Run Versus Artificial Stream
P value without covariate	0.0309	0.0334	0.0130
P value with covariate			
TOC	0.0034	0.0074	0.4538
Turbidity	0.0362	0.0001	0.4345
NO$_3$	0.0408	0.0303	0.0337
Alkalinity	0.0001	0.8271	0.0001
Hardness	0.1470	0.1527	0.0606
Cl	0.0328	0.0387	0.0477
SO$_4$	0.0340	0.0115	0.0120

complex. It is influenced by the metals composition, the absolute concentration, and the test species.

Watenpaugh and Beitinger [27] have demonstrated that fathead minnows do not avoid selenate ions. Investigations with golden shiners *(Notemigonus crysoleucas)* indicate that this species does not avoid selenite or cadmium ions [28]. The results with cadmium are particularly important in light of the substantial increase in toxicity to fathead minnows of the five-metal blend, which included cadmium at very low levels (Table 8). Elements that stimulate a behavioral response at the lowest levels may not be the most important for toxicological considerations. Conversely, highly repulsive yet relatively less toxic concentrations of pollutants may be more environmentally damaging from a habitat suitability standpoint. Environmental impacts cannot be predicted from single-chemical toxicity bioassays, as has been suggested by some authors [29], unless errors of one or two orders of magnitude are considered acceptable.

The response of fish acclimated to the blend was in marked contrast to that of the controls. Fish acclimated to the low dose of the metals blend were completely unresponsive to elevated metals levels. Fish acclimated to the high dose in the laboratory demonstrated a time-dependent response. The response was reversed from initial strong attraction (summer) to mild avoidance (fall) to indifference (winter). The low-dose fish were indifferent to elevated levels after only three months of exposure. This strongly suggests that there is a long-term process of sensory acclimation to metals and the rate and degree of acclimation are dose dependent.

Behavioral acclimation to the high levels of metals may result in fluctuating responses until stable acclimation is reached. What influence the initial high-dose levels (5 and 10% fly-ash slurry), including cadmium, may have had on the response cannot be evaluated with the present data. However, the nature of the effect would presumably be the same in both groups and differ only in degree. Other researchers investigating long-term exposure to metals have observed variations in behavioral and metabolic responses over time. Bengtsson [30] reported a hyperactivity response in the minnow, *Phoxinus phoxinus,* followed by hypoactivity as the zinc concentration and exposure duration increased. Scarfe et al. [31] demonstrated that selected behavioral traits altered by copper exposure were both species specific and reversible three weeks after the cessation of copper exposure. The acclimation of fish to low levels of copper for an extended duration may also affect other behavioral traits, such as feeding orientation and overall activity [31–33]. The ultimate acclimation time may be quite long. McKim et al. [34] and Benoit et al. [10] reported that up to 20 weeks of exposure to a single metal was required to achieve a steady-state body burden.

TABLE 8—*Fathead minnow 96-h LC_{50} and fiducial limits for a blend of four or five metals, expressed as the percentage concentration of the fly-ash basin influent concentration, and calculated metals levels.*

Exposure Group	96-h LC_{50}, %	Fiducial Limits, %	Metal Concentration, mg/L				
			Copper	Chromium	Arsenic	Selenium	Cadmium
Laboratory control	156.8	130.7 to 203.0	2.038	1.098	3.763	0.784	...
Laboratory exposed	195.3	132.9 to 299.3	2.539	1.367	4.687	0.977	...
Field control	147.0	123.3 to 186.2	1.911	1.029	3.524	0.735	...
Field exposed	207.8	167.6 to 251.4	2.701	1.455	4.987	1.093	...
Laboratory controls (5 metals)	80.0	71.2 to 94.7	1.04	0.56	1.92	0.40	0.02

Field Avoidance

The difference in response levels of the control fish illustrates the importance of variation in the background water quality. The fish had significantly different response thresholds to the metals, depending on where they were tested. Fish tested in New River water in summer avoided half the concentration of fish tested in spring (Table 4). Conversely, the avoidance concentration in Adair Run was equivalent to the avoidance level found in the New River in the spring season. The laboratory study demonstrates that variable avoidance response is not a seasonal effect inherent in the fish. The most obvious similarities in water quality parameters in the New River in the spring and Adair Run in summer are in the elevated turbidity and hardness (Table 5). Spring rains and snowmelt increase the amount of material carried by the New River. Adair Run has a small drainage basin and a much steeper slope, with many riffles and waterfalls, so the water velocity is higher and the stream's load is higher into the summer months. Whether the increased sediment load affects the metals directly—for example, by sorption onto clay particles [35], thereby lowering their availability to the fishes' sensory systems—or directly affects the fishes' ability to detect the metals cannot be determined. Other chemical interactions are possible, such as chelation by humic substances, precipitation, and inorganic complexation. Determination of the chemical state of the metals was beyond the scope of this study; however, it is a reasonable hypothesis that the chemical state of the metals may affect how well the fish can sense them. Also, the chemical matrix in which the fish are presented the metals may affect what the fish will choose to respond to. Collins [36] has demonstrated competitive behavioral interactions, due to the physical and chemical characteristics of the water, in migrating Alosids. Sprague et al. [37] reported that Atlantic salmon avoid copper and zinc blends at a level of 0.02 times their incipient lethal level in laboratory tests, but, based on counting fence returns, over ten times that level is required to repulse them in the wild. In the present experiments, the only parameter that significantly affected avoidance in all three test systems was hardness.

Hardness has long been recognized as an important parameter in metals toxicity to fish [38]. The mechanisms by which this occurs are still debated. One view holds that metals are subject to complexation or precipitation, or both, in hard water [39]. Another hypothesis is that calcium reduces membrane permeability to metals [40]. Chapman and McCrady [41] have shown that the toxicity of copper ions varies with pH, as well as with the degree of carbonate complexation. Interpretation of how these toxicological interactions relate to behavioral responses would be speculative at the present time. It is known that metals can harm the olfactory tissue in fish [42,43], but it is not definitely known which sensory system fish use to detect metals. Bodznick [44] has shown that the presence of calcium ions evokes neurological responses in the olfactory tissue of sockeye salmon *(Oncorhynchus nerka)*. Whether the olfactory tissue is sensitive to specific ions or to cations in general is not known.

The physical characteristics of the environment may influence what concentration of metals will be avoided. The slope of the response line from the summer artificial stream trials is as steep as the slope from the laboratory control trials. Fish usually preferred one side of the avoidance enclosure or the other during the control period in the field trials. Because of their initial preference for one side of the chamber, fish were still present in the dosed side at concentrations above those avoided in the laboratory. The concentration in which the fish in the summer artificial stream were avoiding the metals half of the time was 0.7% (34 μg/L metals), which is essentially the same as the laboratory result and is 0.5% of their 96-h LC$_{50}$. However, at that concentration, the laboratory fish occupied the dosed side of the chamber 25% of the time. The field-tested fish did not occupy the dosed side 25% of the time until the concentration was up to 1% (49 μg/L metals) (Fig. 11). The nature of the response was the same, but the actual concentration of metals which drove the fish to the other side was higher.

Turbulence, shadows, proximity to the stream bank, and wild animals in the stream may influence where the fish preferred to be in the Adair Run trials. No predators were seen during testing, but insects, small crayfish, and other fish were usually observed around the enclosure

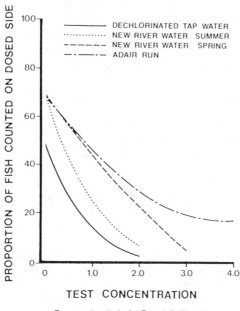

TEST CONCENTRATION

Percent of Ash Pond Influent

FIG. 11—*Avoidance responses of unexposed fathead minnows from all three test systems in their respective water types. The line labeled dechlorinated tap water represents the pooled responses of the four laboratory control groups. The lines are quadratic regressions of the residence time versus concentration.*

during the Adair Run trials. Fish distributions may be influenced by the interactions of cover, substrate type and color, conspecific fish, turbulence, and depth [45]. The fish in the Adair Run and artificial stream spring trial avoided concentrations of about 1% of the 96-h LC_{50}, but did not reach 25% occupancy of the dosed side until the concentrations reached 2.4 and 1.9% (118 μg/L and 93 μg/L), respectively (Fig. 11). Thus, as much as 25% of a population of fathead minnows may not avoid metals concentrations which, if they remain long enough, may result in a complete loss of response to vastly higher levels.

Ishio [46] and Collins [36] have noted that sensitivity and behavioral responses are influenced by both the steepness of the gradient and the absolute value of the stimulus. In the present experiments, the gradient was very steep and offered the fish a clear choice. This situation would be expected in the vicinity of a plume in a river, lake, or estuary. In well-mixed river reaches, the gradient would be more gradual and avoidance response may not be as clearly defined.

Acclimated Fish Avoidance

The response of the fish which had been acclimated to the metals blend for three months was completely different from that of control fish. Acclimated fish did not respond to elevated metals levels in either the artificial stream or in Adair Run (Figs. 9 and 10). Long-term acclimation to very low levels of metals has a profound effect on behavioral responses to those metals. The exposure level of 1.8% (88 μg/L metals) of the fly-ash basin influent concentration was within the range of 50% avoidance by the control fish in the spring artificial stream and Adair Run control trials. The implication is that under certain seasonal conditions, fish may not avoid low levels of metals and may become acclimated to them. When they encounter much higher and possibly harmful levels, they may fail to avoid these levels. Metals-acclimated fish tested in the artificial stream showed no avoidance at concentrations as high as 30% (1470 μg/L metals) of the fly-ash basin influent concentration. The Adair Run trials went as high as 60% (2940 μg/L metals). These levels are 14.4 and 28.8% of the 96-h LC_{50} for these fish. At this concentration, copper alone would be acutely toxic to tolerant fathead minnow populations [47].

Laboratory Versus Field Behavior

The response of fish acclimated to the metals blend for three months in the field is in marked contrast to that of laboratory acclimated and tested fish (Table 4). In the laboratory tests, fatheads acclimated to the high concentration for three months were attracted to elevated levels of the metals blend. Acclimation for longer periods and to lower levels resulted in loss of responsiveness in the laboratory. Sprague et al. [37] noted considerable differences between laboratory metals avoidance by Atlantic salmon and their field distributions. Cairns et al. [48] found that laboratory avoidance and field distributions of fish relative to thermal and chlorine discharges were in close agreement. However, the laboratory testing used thermally acclimated fish, but not chlorine-exposed fish.

This study demonstrates that the difference between laboratory and field tests with metals-acclimated fish is a matter of the time course of the change in behavior. Predictive responses based on short-term laboratory exposure may be erroneous, depending on the exposure level. Predictive results based on unexposed fish may be in closer agreement but will probably overestimate the responsiveness of fish to metals pollution in the wild.

Toxicity

The bioassays clearly demonstrate that when fatheads are exposed to metals, their tolerance to metal toxicity increases. This phenomenon has been observed in fathead minnows and a variety of other fish species [47,49,50]. Increased levels of metallothionein as a result of exposure to metals has been demonstrated in fathead minnows [47]. The rate of metallothionein production and loss requires up to four weeks for equilibrium to be reached [51]. However, changes in resistance as a result of changes in exposure level can be observed in as little as seven days in fathead minnows [47].

Experimental field exposure of fish to metals is not as well studied as laboratory exposures. In the present case, the fish were acclimated for up to nine months to the blend of metals in both laboratory and river water. The resulting increase in tolerance was the same as that when the LC_{50}s were determined in the laboratory. Similar results have been reported for fathead minnows using a blend of copper, cadmium, and zinc [47]. Conversely, when trout are exposed to a blend of zinc, copper, and cadmium in a ratio of 400:20:1 (154 μg/L metals) in the field and also tested for resistance using river water, there is not an apparent increase in LC_{50} values, even though there is an increase in hepatic metallothionein [52]. This is a result of the exposure matrix and not a species difference, since the researchers have previously shown that laboratory exposure to metals results in higher laboratory LC_{50}s [50]. As in the case of the field avoidance tests, the experimental acclimation matrix may have significant impact on the bioassay test results.

Summary and Conclusions

Acclimation to ~90 μg/L of the metals blend in the laboratory and the field results in an increased resistance of 1.25 to 1.41 times that of fish unexposed to metals when tested in the laboratory. The increased resistance does not match the degree of loss of behavioral responsiveness to metals. Addition of cadmium to the blend reduces the 96-h LC_{50} of unacclimated fish to the blend by one half.

Prior exposure to metal blends has a profound effect on fish avoidance behavior. The effect is both dose dependent and time dependent. Unexposed fish avoid ~ 29 μg/L of total metals in the laboratory. This response is consistent over a twelve-month period. Acclimation to ~ 45 μg/L for three months results in a loss of responsiveness to elevated metals levels (~ 245 μg/L) in the laboratory. Acclimation to ~ 90 μg/L for three months results in a preference for elevated metals levels (~ 245 μg/L) in the laboratory. Acclimation for six and nine months to ~ 90 μg/L results in a loss of responsiveness to elevated metals levels (~ 980 μg/L) in the laboratory.

Unexposed fish avoided ~ 71 and ~ 34 μg/L of total metals in an artificial stream during spring and summer conditions, respectively. Acclimated fish did not respond to elevated metals levels (~ 1470 μg/L) in the artificial stream during summer conditions. Unexposed fish avoided ~ 74 μg/L of total metals in a natural stream during summer conditions. Acclimated fish were unresponsive to ~ 2940 μg/L of total metals in a natural stream during summer conditions.

The nature of the response of unexposed fish to the metals blend in the field trials was similar to that of fish in the laboratory trials, but the degree of response was not. This was due to specific chemical differences between the test systems and to the physical settings of the test systems.

Thus, laboratory-derived avoidance data are representative of field-derived data, but may be inaccurate because of differences in the background chemical matrix and the physical setting, and may be entirely erroneous depending upon the acclimation history of the fish. Fly-ash basins and landfills must be operated bearing in mind not only with the percentage of metals removed but also the degree of dilution in the receiving stream.

Acknowledgments

This research was supported by the American Electric Power Co., Columbus, Ohio. Mr. and Mrs. C. Cookingham kindly allowed access onto their property for the in-stream testing.

References

[1] Brungs, W. A. and Mount, D. I., "Introduction to a Discussion of the Use of Aquatic Toxicology Tests for Evaluation of the Effects of Toxic Substances," *Estimating the Hazard of Chemical Substances to Aquatic Life, ASTM STP 657,* J. Cairns, Jr., K. L. Dickson, and A. W. Maki, Eds., American Society for Testing and Materials, Philadelphia, 1978, pp. 15–26.

[2] "Estimating Concern Levels for Concentrations of Chemical Substances in the Environment," unpublished document available from Environmental Effects Branch, Health and Environmental Review Division, Office of Toxic Substances, U.S. Environmental Protection Agency, Washington, DC, 1984.

[3] Mount, D. I. and Stephan, C. E., "A Method for Establishing Acceptable Limits for Fish—Malathion and Butoxyethanol Ester of 2,4-D," *Transactions of the American Fisheries Society,* Vol. 96, No. 2, 1967, pp. 185–193.

[4] Mount, D. I., "Chronic Toxicity of Copper to Fathead Minnows *(Pimephales promelas, Rafinesque),*" *Water Research,* Vol. 2, 1968, pp. 215–223.

[5] White, J. C. and Angelovic, J. W., "Interactions of Chronic Gamma Radiation, Salinity and Temperature on the Morphology of Postlarval Pinfish, *Lagodon rhomboides,*" *Proceedings,* Workshop on Egg, Larval, and Juvenile Stages of Fish in Atlantic Coast Estuaries, A. L. Pacheco, Ed., U.S. NMFS Technical Publication No. 1, Middle Atlantic Coastal Fisheries Center, Highlands, NJ, 1973.

[6] Gardner, G. R. and LaRoche, G., "Copper Induced Lesions in Estuarine Teleosts," *Journal of the Fisheries Research Board of Canada,* Vol. 30, 1973, pp. 363–368.

[7] Servizi, J. A. and Martens, D. W., "Effects of Selected Heavy Metals on Early Life History of Sockeye and Pink Salmon," Progress Report No. 39, International Pacific Salmon Fisheries Community, New Westminster, British Columbia, Canada, 1978.

[8] George, J. D., "Sublethal Effects on Living Organisms," *Marine Pollution Bulletin,* Vol. 1, No. 7, 1970, pp. 107–109.

[9] Eaton, J. G., "Chronic Toxicity for a Copper, Cadmium, and Zinc Mixture to the Fathead Minnow *(Pimephales promelas* Rafinesque)," *Water Research,* Vol. 7, 1973, pp. 1723-1736.

[10] Benoit, D. A., Leonard, E. N., Christensen, G. M., and Fiandt, J. T., "Toxic Effects of Cadmium on Three Generations of Brook Trout *(Salvelinus fontinalis),*" *Transactions of the American Fisheries Society,* Vol. 105, No. 4, 1976, pp. 550–560.

[11] Knittel, M. D., "Heavy Metal Stress and Increased Susceptibility of Steelhead Trout *(Salmo gairdneri)* to *Yersinia ruckeri* infection," *Aquatic Toxicology: Third Symposium, ASTM STP 707,* J. G. Eaton, P. R. Parrish, and A. C. Hendricks, Eds., American Society for Testing and Materials, Philadelphia, 1980, pp. 321-327.

[12] Cairns, J., Jr., "Quantification of Biological Integrity," *The Integrity of Water,* R. K. Balentine and L. J. Guarraic, Eds., EPA 055-001-01068-1, U.S. Environmental Protection Agency, Washington, DC, 1977.

[13] Hartwell, S. I., Cherry, D. S., and Cairns, J., Jr., "Avoidance Response of Schooling Fathead Minnows *(Pimephales promelas)* to a Blend of Metals Before and During a 9-Month Exposure," *Environmental Contamination and Toxicology,* Vol. 6, No. 3, 1987, pp. 177-187.

[14] Cherry, D. S., Larrick, S. R., Cairns, J., Jr., Van Hassel, J., Stetler, D. A., and Ribbe, P. H., "Continuation of Field and Laboratory Studies at the Glen Lyn Plant—pH and Coal Ash Discharges," University Center for Environmental Studies, Virginia Polytechnic Institute and State University, Blacksburg, VA, 1982.

[15] *Methods for Chemical Analysis of Water and Wastes,* EPA 600/4-79-020, U.S. Environmental Protection Agency, Washington, DC, 1979.

[16] *Handbook for Analytical Quality Control in Water and Wastewater Laboratories,* EPA 600/4-79-019, U.S. Environmental Protection Agency, Washington, DC, 1979.

[17] Hendricks, A. C., Dickson K. L., and Cairns, J., Jr., "A Mobile Fish Bioassay Unit for Effluent Testing," report for Manufacturing Chemists Association, University Center for Environmental Studies, Virginia Polytechnic Institute and State University, Blackburg, VA, 1977.

[18] *Methods for Measuring the Acute Toxicity of Effluents to Freshwater and Marine Organisms,* EPA 600/4-85-013, U.S. Environmental Protection Agency, Washington, DC, 1985.

[19] Hartwell, S. I., Jin, H. J., Cherry, D. S., and Cairns, J., Jr., "Evaluation of Statistical Methods for Avoidance Data of Schooling Fish," *Hydrobiologia,* Vol. 131, 1986, pp. 63-76.

[20] *SAS User's Guide: Statistics,* SAS Institute, Inc., Cary, NC, 1982.

[21] Giattina, J. D., Garton, R. R., and Stevens, D. G., "Avoidance of Copper and Nickel by Rainbow Trout as Monitored by a Computer-Based Data Acquisition System," *Transactions of the American Fisheries Society,* Vol. 11, No. 4, 1982, pp. 491-504.

[22] Sprague, J. B., "Avoidance Reactions of Rainbow Trout to Zinc Sulfate Solutions," *Water Research,* Vol. 2, 1968, pp. 367-372.

[23] Kleerekoper, H., Waxman, J. B., and Matis, J., "Interaction of Temperature and Copper Ions as Orienting Stimuli in the Locomotor Behavior of the Goldfish *(Carassius auratus),*" *Journal of the Fisheries Research Board of Canada,* Vol. 30, 1973, pp. 725-728.

[24] Folmar, L. C., "Overt Avoidance Reaction of Rainbow Trout Fry to Nine Herbicides," *Bulletin of Environmental Contamination and Toxicology,* Vol. 15, No. 5, 1976, pp. 509-514.

[25] Black, J. A. and Birge, W. J., "An Avoidance Response Bioassay for Aquatic Pollutants," Report 123, Water Resources Institute, University of Kentucky, Lexington, KY, 1980.

[26] Sprague, J. B., "Avoidance of Copper-Zinc Solutions by Young Salmon in the Laboratory," *Journal of the Water Pollution Control Federation,* Vol. 36, No. 8, 1964, pp. 990-1004.

[27] Watenpaugh, D. E. and Beitinger, T. L., "Absence of Selenate Avoidance by Fathead Minnows *(Pimephales promelas),*" *Water Research,* Vol. 19, 1985, pp. 923-926.

[28] Hartwell, S. I., Jin, H. J., Cherry, D. S., and Cairns, J., Jr., "Toxicity Versus Avoidance Response of Golden Shiners *(Notemigonus crysoleucas)* to Heavy Metals," *Journal of Fish Biology,* in press.

[29] Sloof, W., VanOers, J. A. M., and DeZwart, D., "Margins of Uncertainty in Ecotoxicological Hazard Assessment," *Environmental Toxicology and Chemistry,* Vol. 5, No. 9, 1986, pp. 841-852.

[30] Bengtsson, B. E., "Effect of Zinc on the Movement Pattern of the Minnow *Phoxinus phoxinus* L.," *Water Research,* Vol. 8, 1974, pp. 829-833.

[31] Scarfe, A. D., Jones, K. A., Steele, C. W., Kleerekoper, H., and Corbett, M., "Locomotor Behavior of Four Marine Teleosts in Response to Sublethal Copper Exposure," *Aquatic Toxicology,* Vol. 2, 1982, pp. 335-353.

[32] Drummond, R. A., Spoor, W. A., and Olson, G. F., "Some Short-Term Indicators of Sublethal Effects of Copper on Brook Trout," *Salvelinus fontinalis. Journal of the Fisheries Research Board of Canada,* Vol. 30, 1973, pp. 698-701.

[33] Steele, C. W., "Effects of Exposure to Sublethal Copper on the Locomotor Behavior of the Sea Catfish, *Arius felis,*" *Aquatic Toxicology,* Vol. 4, 1983, pp. 83-93.

[34] McKim, J. M., Olson, G. F., Holcombe, G. W., and Hunt, E. P., "Long-Term Effects of Methylmercuric Chloride on Three Generations of Brook Trout *(Salvelinus fontinalis)*: Toxicity, Accumulation, Distribution, and Elimination," *Journal of the Fisheries Research Board of Canada,* Vol. 30, 1976, pp. 698-701.

[35] *Water-Related Environmental Fate of 129 Priority Pollutants,* EPA 440/4-79-029A, U.S. Environmental Protection Agency, Washington, DC, 1979.

[36] Collins, G. B., "Factors Influencing the Orientation of Migrating Anadromous Fish," *U.S. Fish and Wildlife Service Fishery Bulletin*, Vol. 73, 1952, pp. 375-396.
[37] Sprague, J. B., Elson, P. F., and Saunders, R. L., "Sublethal Copper-Zinc Pollution in a Salmon River—A Field and Laboratory Study," *International Journal of Air and Water Pollution*, Vol. 9, 1965, pp. 531-543.
[38] Black, J. A., Roberts, R. F., Johnson, D. M., Minicucci, D. D., Mancy, K. H., and Allen, H. E., "The Significance of Physiochemical Variables in Aquatic Bioassays of Heavy Metals," G. E. Glass, Ed., *Bioassays Techniques and Environmental Chemistry*, Ann Arbor Sciences, Ann Arbor, MI, 1973, pp. 259-275.
[39] Pickering, Q. H. and Henderson, C., "Acute Toxicity of Some Heavy Metals to Different Species of Warm Water Fishes," *International Journal of Air and Water Pollution*, Vol. 10, 1966, pp. 453-463.
[40] Lloyd, R. "The Toxicity of Zinc Sulfate to Rainbow Trout," *Annals of Applied Biology*, Vol. 48, 1960, pp. 84-94.
[41] Chapman, G. A. and McCrady, J. K., "Copper Toxicity: A Question of Form," *Recent Advances in Fish Toxicology*, R. A. Tubb, Ed., EPA 600/3-77-085, U.S. Environmental Protection Agency, Washington, DC, 1977.
[42] Hara, T. J., Law, Y. M. C. and MacDonald, S., "Effects of Mercury and Copper on the Olfactory Response in Rainbow Trout, *Salmo gairdneri*," *Journal of the Fisheries Research Board of Canada*, Vol. 33, 1976, pp. 1568-1573.
[43] Bodammer, J. E., "The Cytopathological Effect of Copper on the Olfactory Organs of Larval Fish *(Pseudopleuronectes americanus* and *Melanogrammus aeglefinus)*," International Council for Exploration of the Sea, C.M./E:46, 1981.
[44] Bodznick, D., "Calcium Ion, and Odorant for Natural Water Discriminations and the Migratory Behavior of Sockeye Salmon," *Journal of Comparative Physiology and Biochemistry A*, Vol. 127, No. 2, 1978, pp. 157-166.
[45] Casterlin, M. E. and Reynolds, W. W., "Habitat Selection by Juvenile Bluegill Sunfish, *Lepomis machrochirus*," *Hydrobiologia*, Vol. 59, No. 1, 1978, pp. 75-79.
[46] Ishio, S., "Behavior of Fish Exposed to Toxic Substances," O. Jagg, Ed., *Advances in Water Pollution Research*, Vol. 1, Pergamon Press, New York, 1964, pp. 19-40.
[47] Benson, W. H. and Birge, W. J., "Heavy Metal Tolerance and Metallothionein Induction in Fathead Minnows: Results from Field and Laboratory Investigations," *Environmental Toxicology and Chemistry*, Vol. 4, No. 1, 1985.
[48] Cairns, J. Jr., Cherry, D. S., and Giattina, J. D., "Correspondence Between Behavioral Responses of Fish in Laboratory and Field Heated, Chlorinated Effluents," *Proceedings*, International Symposium on Energy and Ecological Modeling, International Society for Ecological Modeling, Louisville, KY, 20-23 April 1981, pp. 207-215.
[49] Dixon, D. G. and Sprague, J. B., "Copper Bioaccumulation and Hepatoprotein Synthesis During Acclimation to Copper by Juvenile Rainbow Trout," *Aquatic Toxicology*, Vol. 1, 1981, pp. 69-81.
[50] Roch, M. and McCarter, J. A., "Hepatic Metallothionein Production and Resistance to Heavy Metals by Rainbow Trout *(Salmo gairdneri)*: I—Exposed to an Artificial Mixture of Zinc, Copper and Cadmium," *Comparative Biochemistry and Physiology*, Vol. 77, No. 1, 1984, pp. 71-75.
[51] McCarter, J. A. and Roch, M., Chronic Exposure of Coho Salmon to Sublethal Concentrations of Copper: III—Kinetics of Metabolism of Metallothionein," *Comparative Biochemistry and Physiology*, Vol. 77, No. 1, 1984, pp. 83-87.
[52] Roch, M. and McCarter, J. A., "Hepatic Metallothionein Production and Resistance to Heavy Metals by Rainbow Trout *(Salmo gairdneri)*: II—Held in a Series of Contaminated Lakes," *Comparative Biochemistry and Physiology*, Vol. 77, No. 1, 1984, pp. 77-82.

Robert J. Livingston[1]

Use of Freshwater Macroinvertebrate Microcosms in the Impact Evaluation of Toxic Wastes

REFERENCE: Livingston, R. J., "Use of Freshwater Macroinvertebrate Microcosms in the Impact Evaluation of Toxic Wastes," *Functional Testing of Aquatic Biota for Estimating Hazards of Chemicals, ASTM STP 988,* J. Cairns, Jr., and J. R. Pratt, Eds., American Society for Testing and Materials, Philadelphia, 1988, pp. 166–218.

ABSTRACT: The author compared the impact of toxic waste sites on three streams in South Carolina and Florida. In addition to the field evaluations of macroinvertebrate distribution (as determined by artificial leaf-pack collectors), a series of acute (4-day) and chronic (30-day) leaf-pack bioassays were carried out with polluted water and leaf-pack microcosms taken from the related study sites. The comparison of laboratory and field data was carried out over a one to two-year period at quarterly intervals. These experiments were designed to evaluate and verify the use of microcosms in predicting field impact and to test for possible seasonality in the laboratory-field relationship.

Leaf-pack data from the field indicated that there was no impact in a South Carolina stream (Myers Creek) because the toxic wastes never entered the surface water system. The four-day leaf-pack bioassay was useful in confirming the lack of adverse effects in the field. Field gradients of macroinvertebrates were evident in Hogtown Creek, Florida (from the low dissolved oxygen and high levels of phenolic derivatives), and in the Little Dry Creek/Dry Creek system, Florida (from the low pH and high lead and aluminum). The acute leaf-pack tests in the Hogtown Creek system did not predict the field impact as well as the chronic tests, whereas both forms of testing in the Little Dry Creek/Dry Creek system were effective in the prediction of field impact. In both systems, the presence of pollution-resistant species complicated the direct extrapolation of bioassay results to field conditions. There was considerable variation in the responses of individual species to the laboratory conditions; however, species richness was a relatively robust laboratory indicator of impact in both streams. Variables that had some influence on the ability to predict field impact using bioassay tests included local features of the stream system, the type of toxic waste, the duration of laboratory exposure to the toxic agents, and the community index used for the comparison. In some instances, the period of sampling was a factor in the laboratory-field relationship. These test results indicate that calibration of the macroinvertebrate microcosm with field data is necessary if such tests are to be used to predict field impact of toxic wastes. Such microcosms, properly used, can be effective for extrapolation of laboratory results to complex field conditions.

KEY WORDS: macroinvertebrates, microcosms, field verification, toxic wastes, bioassays, hazard evaluation

There is a growing need to evaluate the efficacy of laboratory bioassay results in predicting adverse effects of toxic wastes in the field. Does the response of laboratory populations to a toxic substance adequately represent the actual biological response in a receiving system? The vulnerability of a natural system to anthropogenous chemicals is dependent on many factors, physical, chemical, and biological [1]. The diversity, variability, adaptive proclivity, and complexity of

[1]Professor and director, Center for Aquatic Research and Resource Management, Florida State University, Tallahassee, FL 32306.

natural ecosystems preclude simple answers [2]. The extreme range of responses based on inherent differences in aquatic systems adds to such complications. Trophic response is an important factor in such considerations; the need for testing at various levels of biological organization [3] has led to the broader framework of bioassay tests, including the development of multispecies bioassays. Considerable effort has been expended to develop test procedures that are designed to predict and evaluate multispecies responses to toxic substances. Such tests measure a broad range of factors that are considered more closely aligned with ecosystem interactions than tests carried out with single species.

Multispecies tests have various advantages and disadvantages based on the need for sensitivity, simplicity, cost-effectiveness, practicality, and predictability [4]. The use of multispecies testing for comparison, verification, and evaluation of structural and functional parameters of a given ecosystem is generally accepted, although the predictive capability of such methods from system to system remains largely untested [5]. Criteria for verification and extrapolation have been developed [6, 7] but have not been widely applied to real-world situations. Site-specific laboratory/field evaluations are being used with increasing frequency to evaluate toxic waste impact [8]. Elegant comparisons of toxicity tests and the field responses of entire communities have been made to explain covarying gradients of toxic wastes and community variables [9]. However, there are various problems with the verification process, which remains poorly defined. Acute tests may not predict chronic toxicity [10]. Bioassay extrapolations to pond systems may not predict complex, low-level responses of fishes to certain toxicants [11]. Various natural communities do not necessarily respond in the same way to pollutants [12]. In the face of such inconsistencies, it becomes necessary to establish the strengths and weaknesses of specific types of microcosms in the verification of bioassay results with field data.

This paper will address the applicability of bioassay results using freshwater macroinvertebrate microcosms to predict the field impact of toxic wastes on three stream systems.

Methods and Materials

Following a detailed field (habitat) stratification program, a series of sampling stations was located at each of the study sites, which included first- to third-order streams in South Carolina and Florida (Fig. 1). Station placement was based on detailed reviews of existing data and preliminary water chemistry surveys; appropriate (uncontaminated) reference sites were established in the Hogtown Creek and Little Dry Creek/Dry Creek systems in Florida. Detailed station listings are available in previous publications [13]. Rainfall and stream-flow information was provided by the U.S. Forestry Service and the U.S. Geological Survey. Sediment analyses, carried out with core samples (5.0 cm deep and 45 cm^2 in cross-sectional area), included granulometric features and the percentage of organic substances. Dissolved oxygen was measured with a Yellow Springs Instruments Model 57 meter. The pH was measured with a portable Corning 610-A meter calibrated at appropriate intervals with standard solutions. Chemical analyses of metals and organic contaminants in the water and sediments were carried out by Dr. William Cooper (Department of Chemistry, Florida State University, Tallahassee); detailed descriptions of the techniques used have been given elsewhere [13]. Water samples were taken by direct immersion of 1-L glass bottles. "Push-down" corers were used to take sediment chemistry samples (polyvinyl chloride tubes were for metals, aluminum tubes for organics). Organic contaminants were analyzed with Varian 3700 and Perkin-Elmer Sigma 2000 gas chromatographs. These results were corroborated with a Finnigan 4510 gas chromatography/mass spectrometry (GC/MS) system. Metals were analyzed with a Perkin-Elmer 5000 spectrophotometer equipped with standard nebulizer/burners or graphite furnaces.

Artificial leaf packs were developed to sample field populations of epibenthic invertebrates in the field and to run corresponding acute and chronic bioassays in the laboratory. Each leaf-

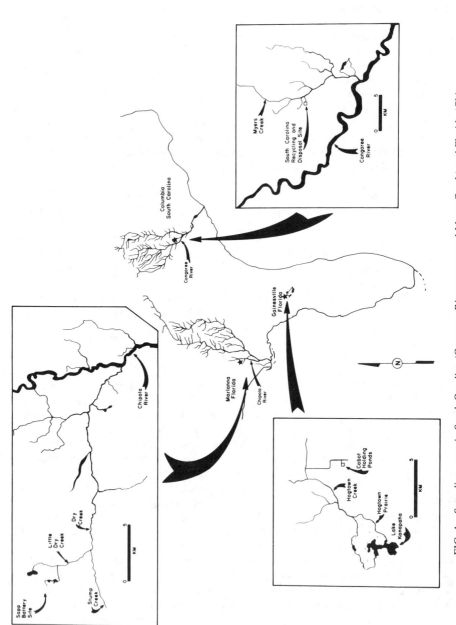

FIG. 1—Sampling areas in South Carolina (Congaree River system and Myers Creek) and Florida (Chipola River system, Little Dry Creek/Dry Creek system, and Hogtown Creek) for verification studies of toxic waste site impacts.

pack collector consisted of 20 Teflon 3.0-cm^2 "leaves" placed inside an 8.0-mm mesh bag. Species accumulation tests with multiple samples at different sites indicated an asymptotic relationship at somewhere between four and five subsamples [13]. Five subsamples were subsequently used in a series of colonization tests. MacArthur-Wilson calculations [14] indicated that equilibrium was reached at between one and two weeks of field exposure. Field transfer experiments supported these results.

Based on the preliminary experiments, a field sampling/laboratory protocol was developed that involved the use of multiple leaf-pack samplers. Field samples (for water and sediment quality and leaf packs) were taken to determine the distribution of organisms along gradients of toxic wastes while laboratory tests were run with microcosms incubated in the field. Field sampling was carried out at monthly intervals in the Little Dry Creek/Dry Creek system (April 1983–April 1985) and at quarterly intervals in Hogtown Creek (July 1984–April 1985). One set of samples was run in Myers Creek (South Carolina) during August 1984. Five samples per station were used after a four-week exposure in the field. The leaf-pack samples were washed through 500-μm sieves; the animals were then preserved and subsequently counted and identified by species.

Samples (100) for the acute (96-h) leaf-pack bioassays (run at the same times as the field transect station sampling) were incubated for approximately four weeks at the reference stations in the related streams. These leaf packs were collected in one-L jars with water from the site. A 96-h bioassay was then run in a temperature-controlled test apparatus. The water temperature in the acute tests was held close to ambient conditions in the field. Tests were run in a greenhouse under ambient light conditions. Ten replicates (two sets of five) were randomly chosen for exposure to the test concentrations. The jar arrangement for the leaf-pack bioassay was random; test solutions at various concentrations (4 L) were replaced each day over the four-day test period without stirring. The test water was taken directly from the contaminated sites; five concentrations (100, 56, 32, 18, and 10%) were maintained by dilution of site water with water taken from the reference stations. In the acute tests, control water (Tables 1a through 1d) was handled in two treatments: one (96S) was changed each day, and the other (96C) was not changed for the duration of the experiment. Chronic tests were run with one laboratory control. Site water was taken at the toxic waste site or in the immediate (adjacent) receiving stream. One field sample (ten leaf packs) was preserved for baseline analysis. Chemical samples (for metals, organics, temperature, pH, and dissolved oxygen) were taken at daily intervals. At the end of each test, all the concentrations were changed to reference site water and left undisturbed for 24 to 36 h to allow differentiation between living and dead animals. Each leaf pack was then harvested in a manner similar to that used for the field samples: the contents of the leaf packs were washed into containers. The animals were dislodged from the mesh bag and Teflon leaves by mechanical agitation. The samples were then washed through 500-μm screens into labeled vials and fixed in 80% ethyl alcohol. All the samples were identified to the species level and counted.

Four-week flow-through tests were run with water taken from Hogtown Creek in an all-glass diluter system [13]. The test waters and dilution waters were taken from the impact sites at three-day intervals. Based on a series of preliminary acute tests, it was determined that low pH was the chief toxic factor in the Little Dry Creek/Dry Creek system; the metal levels comprised an insignificant part of the toxic response of the bioassay organisms. Consequently, the chronic (flow-through) tests for the Little Dry Creek/Dry Creek system were carried out with various (laboratory-controlled) concentrations of pH (metered levels of sulfuric acid) that were comparable to field gradients. All the tests were run at 20°C during the fall months and 24°C during the winter months at ambient light levels. Leaf packs exposed at the reference sites were placed in (one-L) glass jars fitted with 500-μm mesh screens to retain the leaf-pack organisms. Recovery and analysis of the animals were carried out as just described. One-way analysis of variance (ANOVA) was run on the laboratory test results. Comparisons of the various concentrations were made with control 96S; the data analyses were run with and without log transformations.

TABLE 1a—Results of acute leaf-pack bioassays (Hogtown Creek)—27 July 1984.[a]

Result	Field	Control		Concentration of Contaminated Site Water, %				
		96S	96C	100	56	32	18	10
Number of individuals per leaf pack	24.7	23.0	26.3	18.0	29.7	15.1	28.5	35.2[b]
Number of species	34	27	32	24	31	27	29	34
Brillouin diversity	2.68	2.44	2.65	2.27	2.58	2.31	2.46	2.30
Dominant species, number per leaf pack								
Cheumatopsyche sp.	0.5	0.6	1.4	0.3	2.2	0.4	1.4	1.4
Microcylloepus pusillus	4.2	2.7	4.0	5.4	5.6	4.3	6.5	9.4[b]
Planorbidae sp. 1	1.5	1.9	1.6	0.8	1.8	0.8	0.9	0.0[b]
Rheotanytarsus distinctissimus	0.7	5.1	0.4	0.0[b]	0.0[b]	0.0[b]	0.0[b]	0.0[b]
Rheotanytarsus sp.	0.0	0.0	0.9	2.3[b]	3.7[b]	1.4	1.5	6.8[b]
Stenonema smithae	3.0	3.3	4.1	0.0[b]	2.2	2.3	4.9	4.6
Tanytarsus sp.	2.6	0.0	1.6	0.6	2.0[b]	0.4	2.1[b]	2.4[b]

[a]Test conditions:
Dilution water: Station 8
Test water: Station 1
Temperature: 26 ± 2°C
Dissolved oxygen:
At 100% effluent concentration:
Start: 4.9 ppm
End: 7.0 ppm
At 0% effluent concentration:
Start: 4.2 ppm
End: 8.18 ppm
Phenols (total): 4.90 ppm
[b]ANOVA test: $P < 0.05$ (control 96S versus the concentration).

TABLE 1b—Results of acute leaf-pack bioassays (Hogtown Creek)—23 Nov. 1984.[a]

Result	Control			Concentration of Contaminated Site Water, %				
	Field	96S	96C	100	56	32	18	10
Number of individuals per leaf pack	28.6	21.2	48.0	23.6	33.7	31.5	28.2	34.2[b]
Number of species	27	24	27	24	33	29	28	30
Brillouin diversity	2.09	2.02	1.65	1.54	1.92	1.92	1.79	2.21
Dominant species, number per leaf pack								
Caenis sp.	2.7	1.9	1.7	0.9	1.3	1.8	1.1	1.6
Dero sp.	0.2	9.4	29.7	15.0	18.0	16.0	15.6	14.8
Laevapex fuscus	1.4	0.9	1.7	0.7	1.1	0.8	2.1	1.1
Microcylloepus pusillus	10.0	2.0	2.4	9.0	2.9	2.0	1.9	1.9
Tanytarsus sp.	6.4	0.4	1.7	0.6	0.6	0.5	0.2	1.2[b]

[a]Test conditions:
Dilution water: Marine Laboratory well
Test water: Station 1
Temperature: 23 ± 2°C
Dissolved oxygen:
    At 100% effluent concentration:
        Start: 3.6 ppm
        End: 3.3 ppm
    At 0% effluent concentration:
        Start: 7.1 ppm
        End: 7.0 ppm
Phenols (total): 0.26 ppm
[b]ANOVA test: $P < 0.05$ (control 96S versus the concentration).

TABLE 1c—Results of acute leaf-pack bioassays (Hogtown Creek)—20 Jan. 1985.[a]

Result	Control			Concentration of Contaminated Site Water, %				
	Field	96S	96C	100	56	32	18	10
Number of individuals per leaf pack	13.7	32.7	44.9	12.9[b]	16.3	42.7	72.5[b]	58.2[b]
Number of species	25	19	24	9[b]	15[b]	16[b]	26	25
Brillouin diversity	1.98	1.46	1.80	0.55[b]	1.56	0.74	1.078	1.19
Dominant species, number per leaf pack								
Dero sp.	0.2	20.4	20.9	11.2	9.1	35.9	55.5[b]	42.0[b]
Hydroporus sp.	0.1	1.2	1.6	0.7	1.1	1.6	1.0	1.7
Orthocyclops modestus	0.2	1.6	2.8	0.0[b]	0.8	0.5[b]	0.8	0.2[b]
Paratanytarsus sp.	6.6	0.9	6.4	0.0	1.2	1.3	3.4	1.8
Tanytarsus sp.	0.8	2.3	3.3	0.1[b]	0.4[b]	1.1	3.9	5.0[b]

[a]Test conditions:
  Dilution water: Station 8
  Test water: Station 1
  Temperature: 9 ± 2°C
  Dissolved oxygen:
    At 100% effluent concentration:
      Start: 6.0 ppm
      End: 4.2 ppm
    At 0% effluent concentration:
      Start: 9.0 ppm
      End: 11.8 ppm
  Phenols (total): 4.19 ppm
[b]ANOVA test: $P < 0.05$ (control 96S versus the concentration).

TABLE 1d—Results of acute leaf-pack bioassays (Hogtown Creek)—12 April 1985.[a]

Result	Control			Concentration of Contaminated Site Water, %				
	Field	96S	96C	100	56	32	18	10
Number of individuals per leaf pack	57.0	59.9	71.0	10.2[b]	20.4[b]	32.2[b]	33.1[b]	47.0
Number of species	23	26	24	16[b]	19[b]	22	21	25
Brillouin diversity	1.66	2.06	1.90	1.96[b]	1.95	2.25	2.11	2.01
Dominant species, number per leaf pack								
Asellus laticaudatus	1.3	3.2	5.0	0.7	0.4	3.7	3.4	0.9
Dero sp.	0.0	14.6	14.7	3.8[b]	7.4[b]	8.2	9.6	12.4
Hydroporus sp.	1.2	0.6	0.1	0.0	0.0	0.0	0.1	0.2
Orthocyclops modestus	1.9	0.8	0.8	0.5	0.6	0.9	0.4	0.8
Paratanytarsus sp.	32.1	5.1	7.2	0.7[b]	4.7	3.3	3.3	4.5
Tanytarsus sp.	7.7	20.4	28.1	0.0[b]	1.0[b]	5.6[b]	7.6[b]	16.0

[a]Test conditions:
Dilution water: Station 8
Test water: Station 1
Temperature: 18 ± 2°C
Dissolved oxygen:
  At 100% effluent concentration:
    Start: 2.9 ppm
    End: 1.5 ppm
  At 0% effluent concentration:
    Start: 7.0 ppm
    End: 7.0 ppm
Phenols (total): 2.25 ppm
[b]ANOVA test: $P < 0.05$ (control 96S versus the concentration).

Study Area

The two-year study was carried out in three southeastern streams (Myers Creek, South Carolina; Hogtown Creek, Florida; and Little Dry Creek and Dry Creek, Florida) (Fig. 1). A toxic waste site was located on each of these streams. The South Carolina Recycling and Disposal site, originally an acetylene manufacturing plant, was a storage site for various chemical wastes, which reportedly had leaked into the porous soils and groundwater of the Myers Creek area. The waste site is connected to Myers Creek through an intermittent stream. However, field surveys [13] indicated there was no contamination of Myers Creek (in either the water or sediments) by drainage from the waste site. Consequently, studies there were limited to a single field survey.

Hogtown Creek, in Gainesville, Florida, was found to be contaminated by surface discharge from a plant that had distilled pine stumps for production of pine tar products and charcoal from 1945 to 1966. Several lagoons were constructed to hold the wastes, which included pyroligneous acids, higher-molecular-weight alcohols, and phenols. Chemical analysis [13] revealed the presence of a distinct gradient of phenolic derivatives in the water and sediments of Hogtown Creek. This gradient corresponded to levels of low dissolved oxygen along the stream.

Prior to its closing in 1980, a battery-salvage operation at the head of the Little Dry Creek/Dry Creek system, near Marianna, Florida, had released quantities of metals and sulfuric acid into this stream. Distinct gradients of low pH and high lead and aluminum concentrations (leached from the ground by the acidic discharge) were noted in a series of chemical assays [13]. The low pH associated with the discharge had killed major portions of the swamp vegetation in the immediate receiving area. Subsequent sampling at both the Hogtown Creek and Little Dry Creek/Dry Creek sites indicated that the same gradients (low dissolved oxygen, high phenolic derivatives, low pH, and high lead and aluminum) were present throughout the year, although there was considerable variation with the time and surface flow. Seasonal variation was highest in the water chemistry; sediment burdens of pollutants were more stable in time.

Results and Discussion

The pH and metal (lead and aluminum) concentrations in the laboratory treatments of the acute and chronic leaf-pack bioassays were within the range of variation found in the field [13]. Aluminum and lead levels in the water tended to be highest during winter months. The phenolic compounds in Hogtown Creek water showed considerable temporal variability [13]. Laboratory concentrations of phenolic derivatives in the bioassay tests fell within the range of field variation. Dissolved oxygen in the acute bioassays in the laboratory tended to be somewhat higher than in the field, however, although the chronic bioassays replicated field conditions to a considerable degree [13]. The continuous provision of fresh whole effluent for the acute and chronic bioassays was considered to be of great importance. The comparison of the laboratory tests and field distributions of epibenthic invertebrates was facilitated by the simultaneous sampling of field gradients and laboratory responses to bioassay conditions.

Field leaf-pack data in Myers Creek, South Carolina, indicated no direct adverse impact by surface runoff from the toxic waste site. The transect stations were characterized by high numbers of individuals and species during the summer (1984) sampling program. The results of the acute leaf-pack bioassays showed somewhat lower numbers of organisms in the laboratory treatments than in the field. This effect was probably a laboratory artifact based on the loss of organisms during the process of establishing the microcosms. The species richness and diversity indexes were comparable in both the laboratory and the field tests. At all concentrations above the 10% level of effluent taken at the toxic waste site, there was some indication of a low-level impact of water taken from the toxic waste site on the laboratory microcosms; such an effect was based on some alteration of dominance relationships among the macroinvertebrate assemblages. This effect was not noted in the field because such effluents did not actually reach Myers Creek.

Data from the acute leaf-pack tests run with water taken from Hogtown Creek are given in Tables 1a through 1d. Numerical abundance during the July test was somewhat elevated at low concentrations because of increased numbers of various species relative to the controls. This pattern was also apparent during the November experiments. During the January test, there was again an increase in overall numbers at low concentrations; however, species richness (and, to a certain extent, species diversity) were adversely affected because of the elimination of certain species. Recovery of the species richness and diversity indexes was noted in the 18% and 10% treatments. The dominant species varied seasonally in the field, where species-specific differences coincided with the just-mentioned changes in the community indexes. In the April test, a similar pattern of species richness and diversity was noted; however, numerical abundance in this test showed a dose-dependent relationship. Such results tended to coincide with the levels of phenol, with the least overall impact noted during November 1984 (phenols = 0.26 ppm).

Results of chronic, flow-through tests (Tables 2a and 2b) indicated significant increases in numbers at the 50% level and below during the fall period. The species richness and diversity indexes at the top concentration were lower than those in the controls. Populations of Dero sp. and other species were actually favored at various concentrations, usually from 13 to 50%.

The winter test showed a complete loss of the biota at 100% concentrations of water taken at Station 1. At 50% concentrations, the abundance, species richness, and diversity were reduced relative to the controls. At the next two lower concentrations, high abundance was noted (as in

TABLE 2a—*Flow-through (chronic) leaf-pack bioassay results (Hogtown Creek)—October–November 1984.*[a]

Result	Field Control	Concentration of Effluent, %					
		0	100	50	25	13	6.3
Total number	377	122	245	2311[b]	823[b]	441[b]	293[b]
Number of species	31	20	11[b]	21	23	29	22
Brillouin diversity	2.47	2.20	1.07[b]	0.52[b]	1.49	1.99	1.91
Dominant species, total number							
Dero sp.	0	8	167[b]	2076[b]	490[b]	115[b]	119[b]
Microcylloepus pusillus	104	31	38	42	50	47	39
Physella sp.	10	24	8	12	103	128	30
Laevapex fuscus	27	7	1	29	37	65	42
Caenis sp.	52	3	12	25	35	23	14
Orthocyclops modestus	2	0	1	15	39	6	5
Tanytarsus sp.	36	4	0	0	3	2	8
Stenelmis fuscata	11	7	0	1	5	11	5
Asellus laticaudatus	5	1	0	4	8	2	1
Rheotanytarsa sp.	19	0	0	0	0	0	0

[a]Test conditions:
  Test water: Station 1
  Temperature: 24 ± 2°C
  Average dissolved oxygen:
    At 100% effluent: 3.3 ppm
    Control: 8.3 ppm
  Total phenols:
    Week 1: 0.497 ppm
    Week 2: 2.88 ppm
    Week 3: 2.40 ppm
    Week 4: 1.77 ppm
    Week 5: 2.87 ppm
    Week 6: 1.46 ppm
[b]ANOVA test: $P < 0.05$ (control versus the concentration).

TABLE 2b—*Flow-through (chronic) leaf-pack bioassay results (Hogtown Creek)—January–February 1985.*[a]

Result	Field Control	Concentration of Effluent, %					
		0	100	50	25	13	6.3
Total number	243	375	0^b	122	1158^b	721^b	381
Number of species	20	16	0^b	5^b	14	13	13
Brillouin diversity	1.43	1.62	0^b	0.52^b	0.41^b	0.93^b	1.09
Dominant species, total number							
Dero sp.	0	134	0^b	104	1057^b	527^b	268^b
Orthocyclops modestus	7	123	0	11	62	108	34
Paratanytarsus sp.	153	0	0	0	0	0	0
Hydroporus sp.	8	39	0	4	9	32	30
Physella sp.	0	14	0	0	1	26	20
Laevapex fuscus	2	25	0	0	8	16	8
Asellus laticaudatus	2	20	0	0	1	2	10
Corynoneura sp.	26	0	0	0	0	0	0
Microcylloepus pusillus	15	3	0	0	1	0	1
Tanytarsus sp.	6	1	0	0	1	0	0

[a] Test conditions:
Test water: Station 1
Temperature: $20 \pm 2°C$
Average dissolved oxygen:
At 100% effluent: 2.4 ppm
Control: 6.9 ppm
Total phenols:
Week 1: 4.19 ppm
Week 2: 0.31 ppm
Week 3: 4.32 ppm
[b] ANOVA test: $P < 0.05$ (control versus the concentration).

the fall experiment). Species such as *Dero* sp. were favored at moderate to low concentrations. The results of the winter chronic bioassay test closely followed those of the fall chronic test, although there appeared to be a more severe toxic effect during the winter test that could be related to higher levels of phenols and lower dissolved oxygen. These data indicate that, at high concentrations of phenolic derivatives (0.497 to 2.873 ppm in the fall test, 0.312 to 4.323 ppm in the winter) and low oxygen concentrations, epibenthic organisms suffered significant adverse effects at high concentrations, whereas, as the concentrations decreased, certain species actually flourished beyond the control levels. Recovery to control levels occurred at the lowest concentrations. The chronic leaf-pack bioassays run with water from Hogtown Creek thus showed comparable results in the repeated experiments.

A comparison of the field and laboratory data (Fig. 2) indicates that field abundance is reduced at low levels of dissolved oxygen and at certain concentrations of phenolic derivatives in the water. Relatively high numbers are found near phenol levels of 1 ppm during July 1984. This pattern is not always evident in the static leaf-pack tests; the numbers remain high, even at relatively high levels of organic substances, at various times of the year. The chronic tests, on the other hand, showed a pattern similar to that in the field. Species richness patterns tended to indicate a similar relationship between the results of the chronic tests and the field data. The acute tests thus do not predict the field response to the low dissolved oxygen and high levels of phenolic derivatives as well as the chronic, flow-through tests do. The pattern of the distribution of species diversity indexes (Fig. 3) underscores the closer similarity of the field response to the chronic bioassay results than to the acute tests. Species distributions (Fig. 4) show that populations such as *Tanytarsus* sp. appear sensitive to low concentrations of phenolic derivatives and low dissolved oxygen. Once again, the chronic bioassays predict such distributions, whereas the

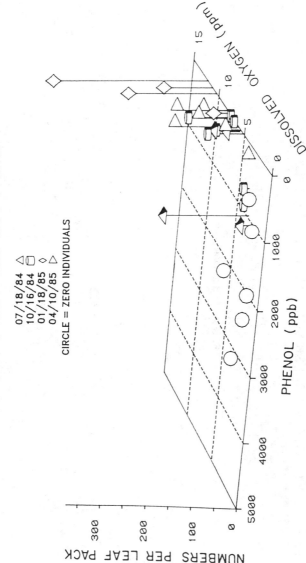

FIG. 2—Comparison of leaf-pack data as a function of dissolved oxygen and phenolic derivatives in waters taken from Hogtown Creek, Gainesville, Florida. The data are presented on a quarterly basis.

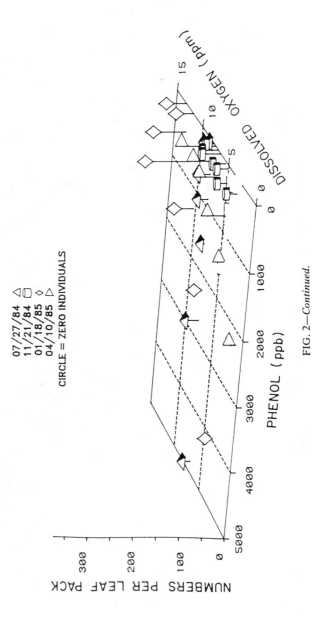

FIG. 2—*Continued.*

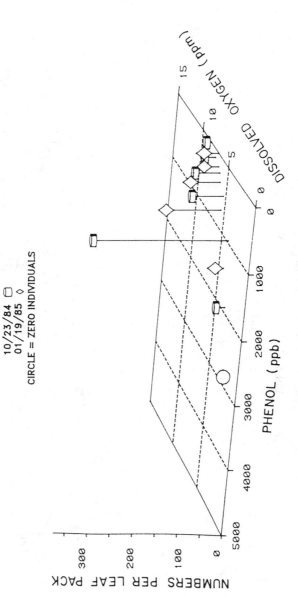

NUMBER OF INDIVIDUALS PER LEAF PACK
HOGTOWN CREEK LEAF PACK FLOW—THROUGH BIOASSAYS

10/23/84
01/19/85

CIRCLE = ZERO INDIVIDUALS

DISSOLVED OXYGEN (ppm)

PHENOL (ppb)

NUMBERS PER LEAF PACK

FIG. 2—*Continued.*

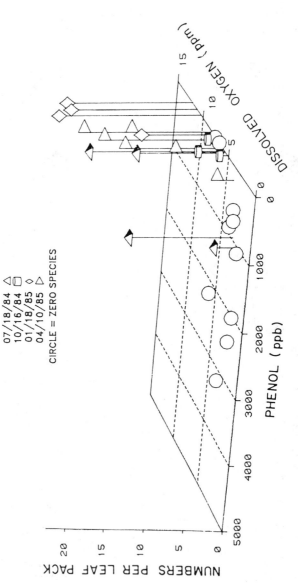

FIG. 2—Continued.

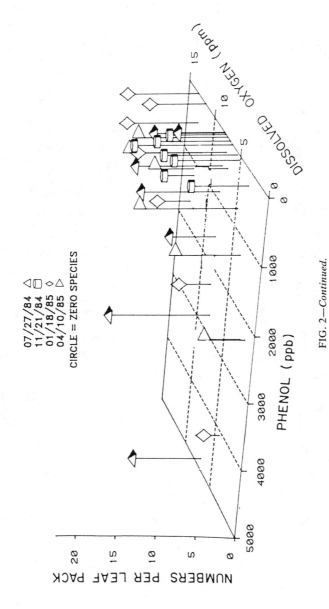

NUMBER OF SPECIES PER LEAF PACK
HOGTOWN CREEK LEAF PACK STATIC BIOASSAYS

07/27/84 △
11/21/84 ▭
01/18/85 ◇
04/10/85 △

CIRCLE = ZERO SPECIES

DISSOLVED OXYGEN (ppm)

PHENOL (ppb)

NUMBERS PER LEAF PACK

FIG. 2—*Continued.*

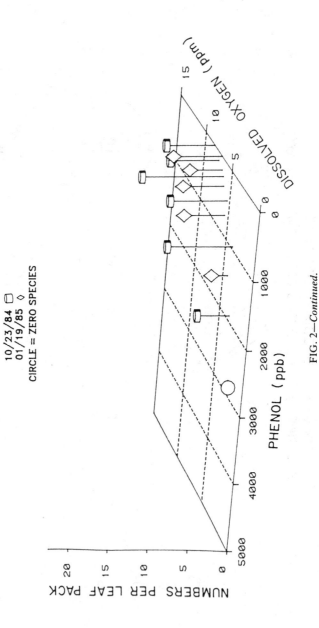

FIG. 2—Continued.

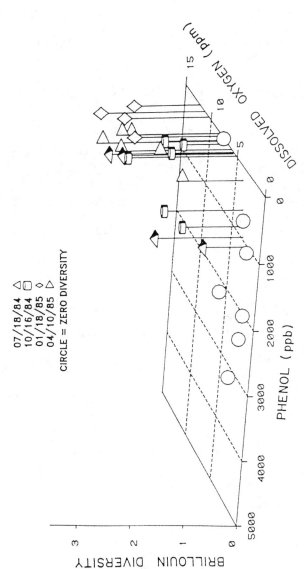

FIG. 3—*Comparison of the Brillouin diversity index as a function of dissolved oxygen and phenolic derivatives in waters taken from Hogtown Creek, Gainesville, Florida. The data are presented on a quarterly basis.*

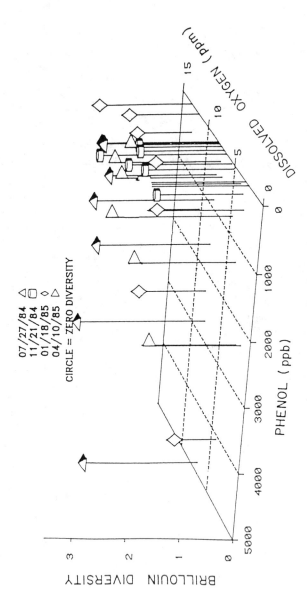

FIG. 3—*Continued.*

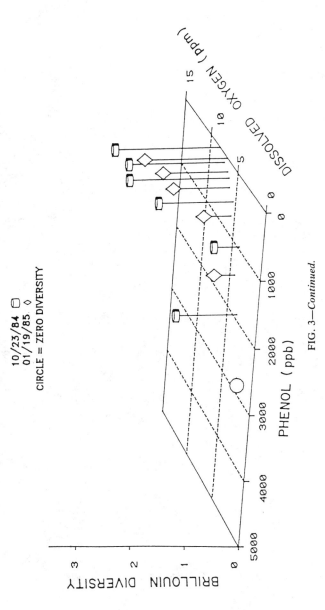

FIG. 3—*Continued.*

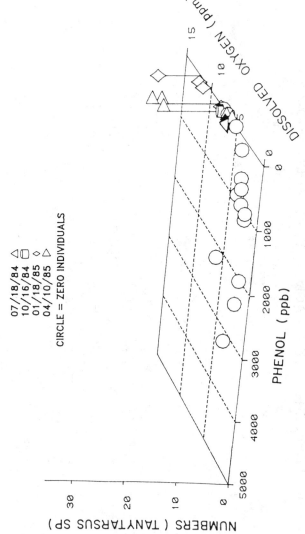

FIG. 4—*Comparison of the response of an individual population as a function of dissolved oxygen and phenolic derivatives in waters taken from Hogtown Creek, Gainesville, Florida. The data are presented on a quarterly basis.*

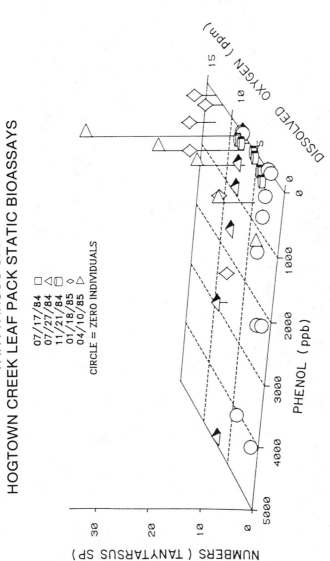

TANYTARSUS SP.
HOGTOWN CREEK LEAF PACK STATIC BIOASSAYS

07/17/84
07/27/84
11/21/84
01/18/85
04/10/85

CIRCLE = ZERO INDIVIDUALS

FIG. 4—*Continued.*

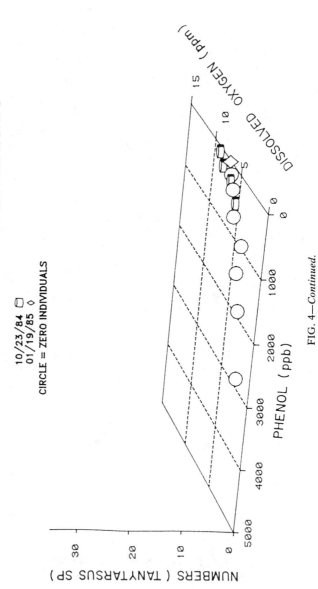

FIG. 4—Continued.

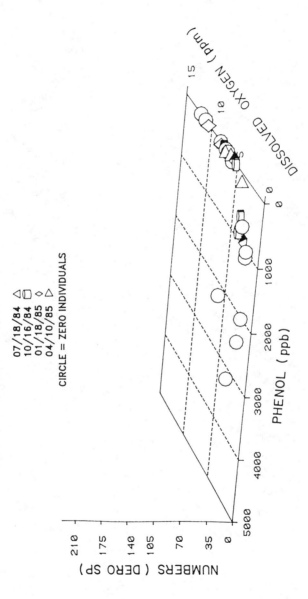

DERO SP.
HOGTOWN CREEK FIELD LEAF PACKS

07/18/84
10/16/84
01/18/85
04/10/85

CIRCLE = ZERO INDIVIDUALS

DISSOLVED OXYGEN (ppm)

PHENOL (ppb)

NUMBERS (DERO SP)

FIG. 4—*Continued*.

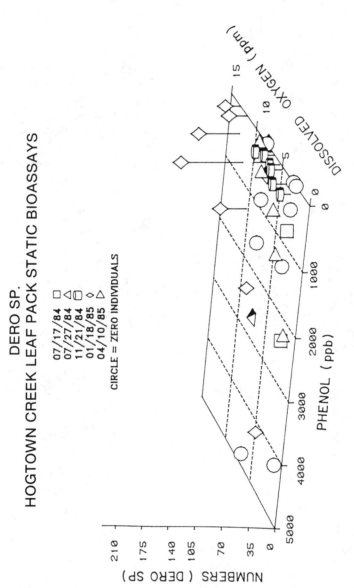

FIG. 4—*Continued.*

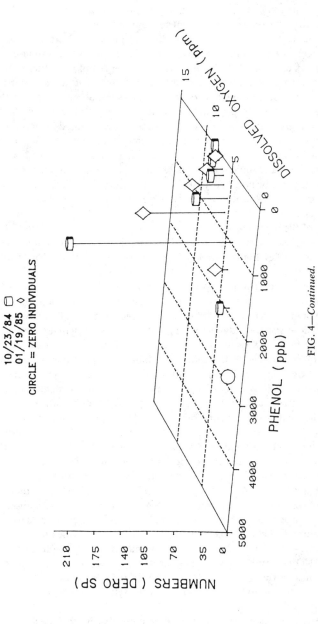

DERO SP.
HOGTOWN CREEK LEAF PACK FLOW—THROUGH BIOASSAYS

10/23/84
01/19/85
CIRCLE = ZERO INDIVIDUALS

FIG. 4—*Continued.*

static tests are less compatible with the field results. The response of species such as *Dero* sp., however, is accentuated in the chronic tests in comparison with the field values: in this case, such partially resistant species are actually favored by long-term exposure in the laboratory to low levels of organic contaminants such as the phenolic derivatives. Such populations are enhanced at moderate concentrations in both the static and chronic tests and are thus not good laboratory indicators of the field response to phenolic derivatives.

Further comparison between the chronic bioassays and field results indicates that, although there are distinct seasonal differences in the response of invertebrate assemblages to low dissolved oxygen and phenolic derivatives, the general pattern of response is similar. In the field, high numbers of insects (*Chironomus decorus* group) and immature tubificid worms showed a pattern of resistance similar to that shown by *Dero* sp. in their winter predominance. Thus, when the enhanced responses of resistant species are accounted for in the chronic bioassays, such tests are adequate predictors of field response. Presumably, the less predictive acute results could be due to the short exposure time, which resulted in an incomplete response to the toxic agents. A comparison of the fall and winter acute and chronic effects (Tables 1b, 1c, 2a, and 2b), indicates that the tendency for increased numbers of individuals in the acute tests was accentuated in the chronic tests. This result illustrates the laboratory artifact in which certain species are favored by laboratory conditions when exposed to low to moderate concentrations of organic contaminants. Species richness appears to indicate a dose-dependent response in both the acute and chronic tests.

Results of the acute leaf-pack tests run with organisms taken from Little Dry Creek and Dry Creek indicate significant seasonal differences in the impact of site runoff on the laboratory microcosms (Tables 3a through 3d). For a comparison with the Hogtown Creek data, only the quarterly tests are presented. During the summer (1984) test, numerical abundance was somewhat reduced because of dose-specific losses of dominant species such as *Hydroptila* and *Tanytarsus*. However, such differences were not statistically significant. Species richness and diversity were not affected by the treatment. In the December (1984) and April (1985) runs, however, there were dose-specific reductions of abundance, species richness, and diversity indexes, which appeared to be adversely affected by the water taken from the toxic waste site. There were reductions of key species in the short-term bioassays; such losses contributed to the observed dose-specific changes in community structure. In each case, such adverse effects were noted at reduced pH levels (less than 3.7). The numbers of organisms and species were low in the 96S control during the January 1985 test; consequently, the statistical analysis showed an inverse response in comparison with the other tests. However, in comparison with the other controls (field, 96C), such numbers were somewhat reduced at the higher concentrations.

Results of the chronic leaf-pack tests (Tables 4a and 4b) run during the fall indicated reductions of numerical abundance and species richness in the exposure to low pH, with recovery at levels ranging from 7.7 to 2.0%. Brillouin diversity followed these trends. Species such as *Stenelmis* sp. did well at high treatment concentrations, whereas other populations *(Paraleptophlebia)* were adversely affected. ANOVA tests confirmed these observations, with only the lowest two concentrations not significantly ($P < 0.05$) different from the laboratory controls in terms of numerical abundance and species richness during the fall test. The winter chronic test showed a pH-dependent reduction of all community indexes, with recovery at the 2.8% concentration. ANOVA results ($P < 0.05$) indicated statistical differences of the top four treatments from controls. Although the winter tests indicated a somewhat more adverse response to low pH, these results confirmed the findings of the fall (1984) bioassay. They were also in general agreement with the results of the acute tests.

Comparisons of the bioassay tests with field results (Fig. 5) indicate that relatively high numbers of organisms are found at various metal (lead and aluminum) concentrations and low pH levels in the field as a result of specific resistant species. Species richness (Fig. 5) and diversity (Fig. 6) tended to be reduced in the field as a function of pH; the static tests confirmed these findings. Species such as *Dero* (Fig. 7) were considerably reduced at low pH levels, although the

TABLE 3a—Results of acute leaf-pack bioassays (Little Dry Creek/Dry Creek system)—18 July 1984.[a]

Result	Control			Concentration of Contaminated Site Water, %				
	Field	96S	96C	100	56	32	18	10
Number of individuals per leaf pack	115.5	98.4	103.1	87.0	64.5	85.5	95.6	...
Number of species	47	40	42	44	38	42	42	...
Brillouin diversity	2.58	2.40	2.44	2.47	2.59	2.50	2.33	...
Dominant species, number per leaf pack								
Crangonyx floridanus	0.7	2.9	2.8	7.0	1.2	3.1	3.0	...
Hydroptila sp.	32.2	26.4	24.4	15.2	11.8	15.1	17.0	...
Lebertia sp.	8.9	17.1	21.2	18.9	13.0	18.3	23.3	...
Lirceus lineatus	2.8	1.2	3.3	3.5	2.1	1.4	1.8	...
Stenelmis sp. (larvae)	13.4	15.9	14.6	14.4	8.2	13.2	21.0	...
Stempellinella sp.	2.1	2.8	3.9	7.8	5.7	10.5[b]	7.2[b]	...
Tanytarsus sp.	16.5	8.7	9.3	1.1[b]	2.1[b]	1.7[b]	1.4[b]	...

[a] Test conditions:
Temperature: 24 ± 2°C
pH:
   At 100% effluent concentration:
      Start: 4.1 pH
      End: 4.3 pH
   At 0% effluent concentration:
      Start: 5.3 pH
      End: 5.4 pH
Metal concentrations:
   Lead: 2.52 ppm
   Aluminum: 0.85 ppm
[b] ANOVA test: $P < 0.05$ (control 96S versus the concentration).

TABLE 3b—Results of acute leaf-pack bioassays (Little Dry Creek/Dry Creek system)—5 Dec. 1984.[a]

Result	Field	Control		Concentration of Contaminated Site Water, %				
		96S	96C	100	56	32	18	10
Number of individuals per leaf pack	28.2	15.4	22.5	5.9[b]	7.3[b]	10.3[b]	7.4[b]	11.4
Number of species	21	26	21	18[b]	17[b]	16[b]	29	20
Brillouin diversity	1.92	2.53	2.39	2.19[b]	2.04[b]	2.09[b]	2.53	2.32
Dominant species, number per leaf pack, *Lebertia* sp.	2.0	1.4	1.2	0.7	1.2	2.2	1.2	1.9

[a]Test conditions:
Temperature: $12 \pm 2°C$
pH:
  At 100% effluent concentration:
    Start: 3.6 pH
    End: 3.7 pH
  At 0% effluent concentration:
    Start: 5.7 pH
    End: 5.6 pH
Metal concentrations:
  Lead: 3.05 ppm
  Aluminum: 4.2 ppm
[b]ANOVA test: $P < 0.05$ (control 96S versus the concentration).

TABLE 3c—Results of acute leaf-pack bioassays (Little Dry Creek/Dry Creek system)—15 Jan. 1985.[a]

Result	Control			Concentration of Contaminated Site Water, %				
	Field	96S	96C	100	56	32	18	10
Number of individuals per leaf pack	16.3	3.3	12.1	5.7	9.2	15.3[b]	10.7[b]	12.0[b]
Number of species	29	12	26	22	25[b]	32[b]	27[b]	28[b]
Brillouin diversity	2.60	1.75	2.40	2.17	2.37	2.66[b]	2.58[b]	2.50[b]
Dominant species, number per leaf pack Tanytarsus sp.	1.6	0.0	2.6	0.3	2.0[b]	2.0[b]	0.9	2.3[b]

[a] Test conditions:
Temperature: $10 \pm 2°C$
pH:
At 100% effluent concentration:
Start: 3.6 pH
End: 3.7 pH
At 0% effluent concentration:
Start: 5.4 pH
End: . . . .
Metal concentrations:
Lead: 3.59 ppm
Aluminum: 8.30 ppm
[b] ANOVA test: $P < 0.05$ (control 96S versus the concentration).

TABLE 3d—Results of acute leaf-pack bioassays (Little Dry Creek/Dry Creek system)—19 April 1985.[a]

Result	Control			Concentration of Contaminated Site Water, %				
	Field	96S	96C	100	56	32	18	10
Number of individuals per leaf pack	20.3	43.7	93.0	3.0[b]	9.2[b]	20.7[b]	20.2[b]	35.8
Number of species	25	28	31	13[b]	23[b]	27	23	28
Brillouin diversity	2.07	1.67	1.96	1.67[b]	2.3[b]	2.02	2.27	2.23
Dominant species, number per leaf pack								
Crangonyx floridanus	7.9	20.4	19.8	0.0[b]	0.8[b]	4.9[b]	4.5[b]	14.0
Eurycercus sp.	2.8	1.9	37.1	0.0[b]	0.1[b]	0.0[b]	0.6[b]	1.3
Orthocyclops modestus	1.6	4.1	3.8	0.1[b]	0.9[b]	1.5[b]	2.4	4.4
Psectrocladius sp.	2.0	1.2	4.0	1.2	1.9	3.8[b]	3.2[b]	2.9[b]
Tanytarsus sp.	0.9	4.0	7.1	0.3[b]	1.4[b]	5.6	1.6	2.7

[a]Test conditions:
Temperature: 21 ± 2°C
pH:
  At 100% effluent concentration:
    Start: 3.2 pH
    End: 3.4 pH
  At 0% effluent concentration:
    Start: 5.2 pH
    End: 5.2 pH
Metal concentrations:
  Lead: 1.61 ppm
  Aluminum: 4.6 ppm
[b]ANOVA test: $P < 0.05$ (control 96S versus the concentration).

TABLE 4a—*Flow-through (chronic) leaf-pack bioassay results (Little Dry Creek/Dry Creek system)—October–November 1984.*[a]

Result	Field Control	Effluent Concentration, %					
		0	100	40	16	7.7	2.0
Total number of individuals	538	462	337[b]	215[b]	255[b]	628	503
Number of species	37	39	23[b]	25[b]	27[b]	38	41
Brillouin diversity	2.47	2.80	1.83[b]	2.25	2.66	2.72	3.01
Dominant species, total number							
Stenelmis sp.	71	50	131[b]	66	43	42	35
Lebertia sp.	93	48	66	15	18	80	73
Paraleptophlebia sp.	67	102	10	3	59	73	84
Lirceus lineatus	15	28	24	24	10	151	19

[a] Test conditions:
  Test water: Station 1
  Temperature: 24 ± 2°C
  Average pH:
    At 100% effluent: 2.55 pH
    Control: 8.1 pH
[b] ANOVA test: $P < 0.05$ (control versus the concentration).

TABLE 4b—*Flow-through (chronic) leaf-pack bioassay results (Little Dry Creek/Dry Creek system)—January–February 1985.*[a]

Result	Field Control	Effluent Concentration, %					
		0	100	40	16	7.7	2.0
Total number of individuals	150	33	2[b]	7[b]	19[b]	16[b]	34
Number of species	25	14	2[b]	6[b]	8[b]	9[b]	16
Brillouin diversity	2.43	1.94	0.35[b]	1.12[b]	1.34	1.52	2.05
Dominant species, total number							
Krenosmittia sp.	30	2	1	0	6	2	1
Lebertia sp.	16	5	0	0	0	0	3
Tanytarsus sp.	18	1	0	0	0	0	1
Cicrotendipes sp.	15	1	0	0	0	1	19
Ablabesmyia sp.	5	4	0	1	0	1	5

[a] Test conditions:
  Test water: Station 1
  Temperature: 20 ± 2°C
  Average pH:
    At 100% effluent: 2.51 pH
    Control: 5.96 pH
[b] ANOVA test: $P < 0.05$ (control versus the concentration).

impact of high metal concentrations (lead and aluminum) was not as clear. However, field numbers of *Tanytarsus* were high in areas of high lead and low pH; such results were not found in the acute tests. Numerical abundance in the field as a function of pH was not necessarily found in the static bioassays because of the field presence of resistant species. However, species richness was a good indicator in both types of laboratory tests (acute and chronic) when compared with the field response of organisms in the Little Dry Creek/Dry Creek system to the toxic waste site. Seasonal differences of such responses could lead to problems of interpretation if testing occurred during the occasional periods of low response.

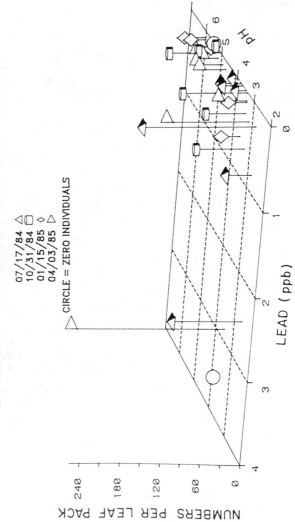

FIG. 5—*Comparison of leaf-pack data as a function of pH and lead or aluminum in waters taken from the Little Dry Creek/Dry Creek system, Marianna, Florida. The data are presented on a quarterly basis.*

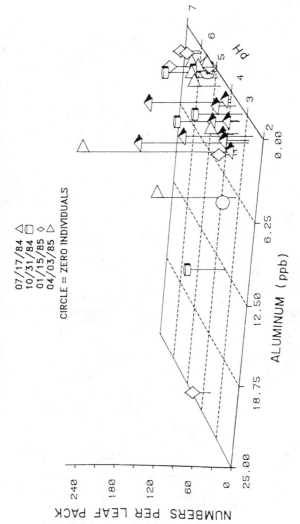

FIG. 5—*Continued.*

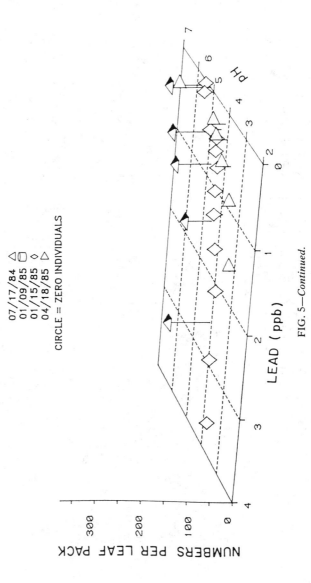

NUMBER OF INDIVIDUALS PER LEAF PACK
LITTLE DRY—DRY CREEK LEAF PACK STATIC BIOASSAYS

07/17/84 △
01/09/85 ⬡
01/15/85 ◇
04/18/85 △

CIRCLE = ZERO INDIVIDUALS

FIG. 5—Continued.

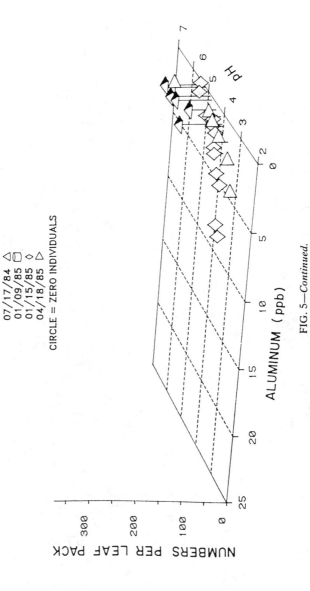

FIG. 5—*Continued.*

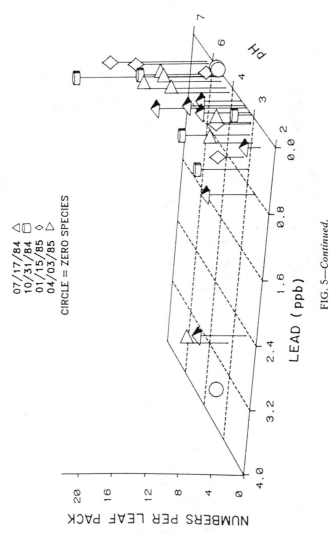

FIG. 5—*Continued.*

NUMBER OF SPECIES PER LEAF PACK
LITTLE DRY—DRY CREEK FIELD LEAF PACKS

07/17/84 △
10/31/84 ⬠
01/15/85 ◇
04/03/85 △

CIRCLE = ZERO SPECIES

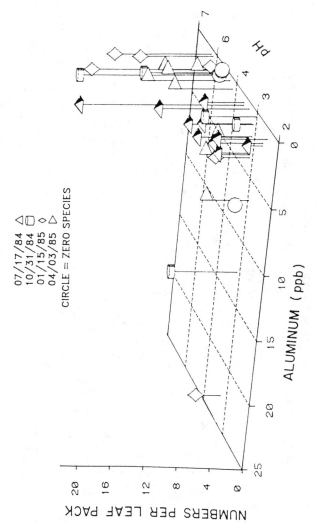

FIG. 5—Continued.

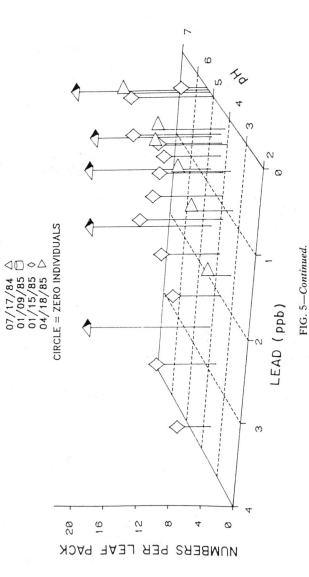

NUMBER OF SPECIES PER LEAF PACK
LITTLE DRY—DRY CREEK LEAF PACK STATIC BIOASSAYS

07/17/84 △
01/09/85 ▢
01/15/85 ◇
04/18/85 △

CIRCLE = ZERO INDIVIDUALS

FIG. 5—Continued.

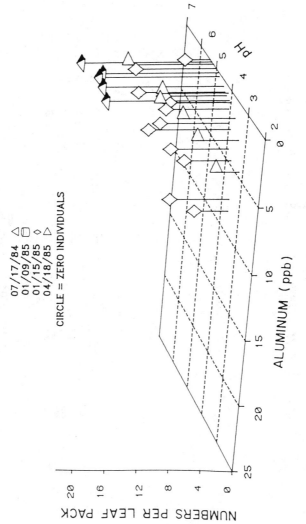

NUMBER OF SPECIES PER LEAF PACK
LITTLE DRY—DRY CREEK LEAF PACK STATIC BIOASSAYS

07/17/84 △
01/09/85 ▢
01/15/85 ◇
04/18/85 △

CIRCLE = ZERO INDIVIDUALS

FIG. 5—*Continued.*

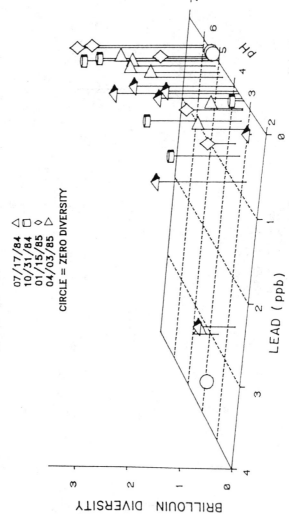

FIG. 6—*Comparison of the Brillouin diversity index as a function of pH and lead or aluminum in waters taken from the Little Dry Creek/Dry Creek system, Marianna, Florida. The data are presented on a quarterly basis.*

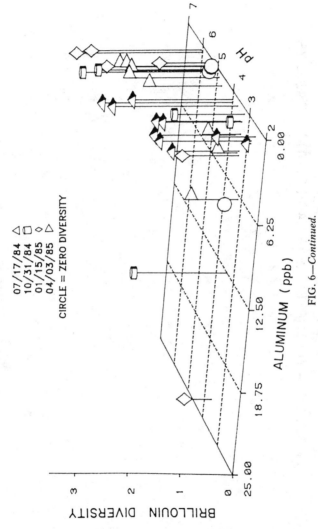

BRILLOUIN DIVERSITY
LITTLE DRY—DRY CREEK FIELD LEAF PACKS

07/17/84
10/31/84
01/15/85
04/03/85

CIRCLE = ZERO DIVERSITY

FIG. 6—*Continued.*

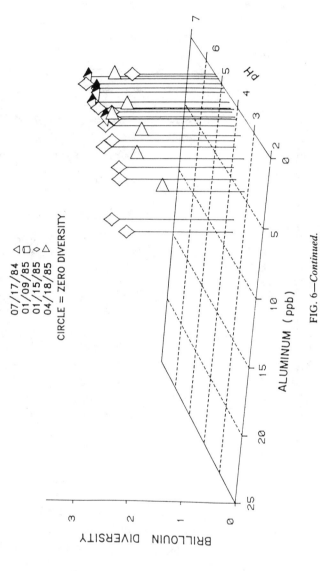

FIG. 6—*Continued.*

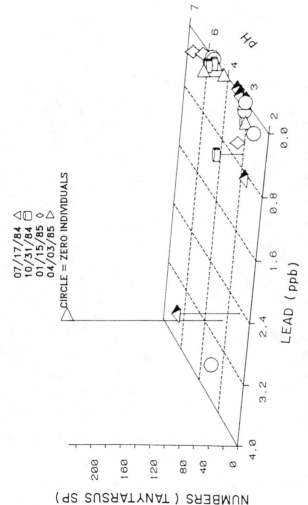

FIG. 7—*Comparison of the response of an individual population as a function of pH and lead or aluminum in waters taken from the Little Dry Creek/Dry Creek, Marianna, Florida. The data are presented on a quarterly basis.*

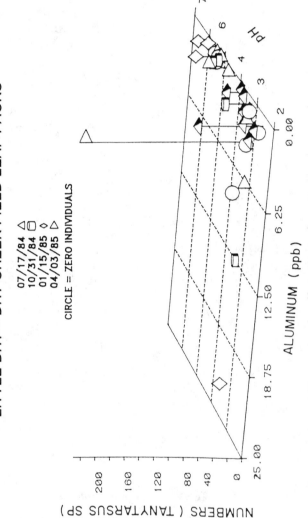

FIG. 7—*Continued.*

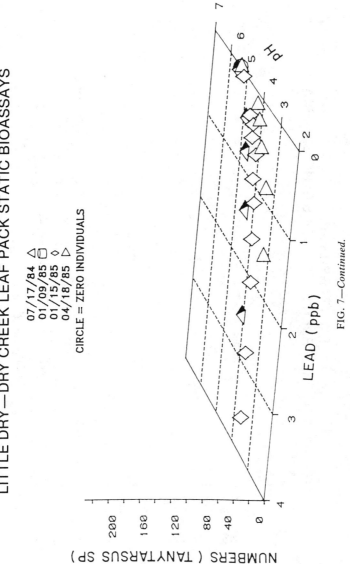

TANYTARSUS SP.
LITTLE DRY—DRY CREEK LEAF PACK STATIC BIOASSAYS

07/17/84
01/09/85
01/15/85
04/18/85

CIRCLE = ZERO INDIVIDUALS

NUMBERS (TANYTARSUS SP)

LEAD (ppb)

pH

FIG. 7—Continued.

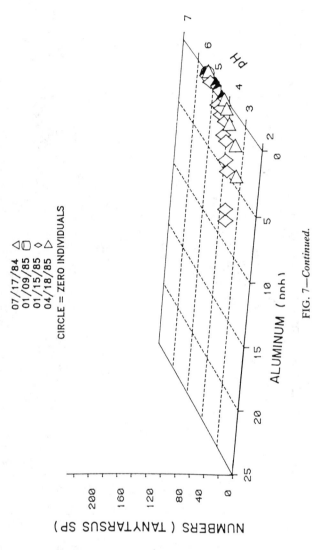

FIG. 7—*Continued.*

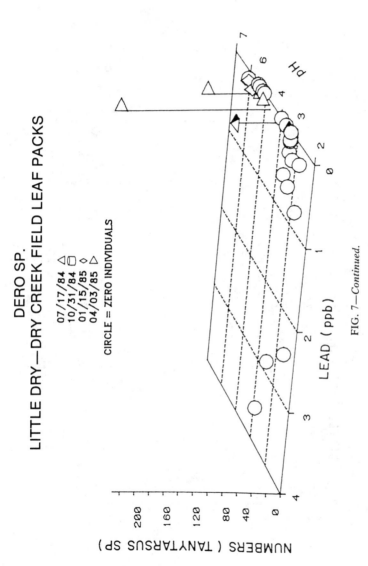

FIG. 7—*Continued.*

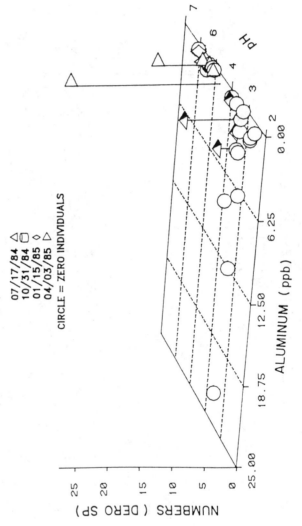

DERO SP.
LITTLE DRY—DRY CREEK FIELD LEAF PACKS

07/17/84
10/31/84
01/15/85
04/03/85

CIRCLE = ZERO INDIVIDUALS

FIG. 7—*Continued*.

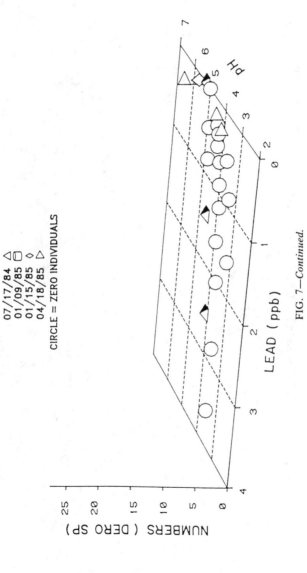

DERO SP.
LITTLE DRY—DRY CREEK LEAF PACK STATIC BIOASSAYS

07/17/84
01/09/85
01/15/85
04/18/85

CIRCLE = ZERO INDIVIDUALS

FIG. 7—*Continued.*

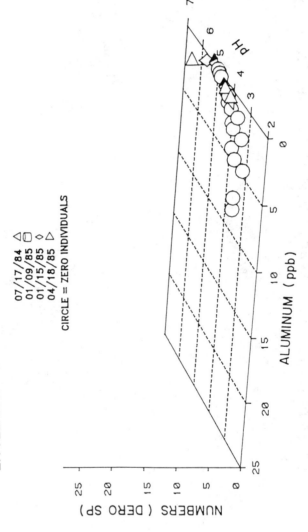

FIG. 7—Continued.

These comparisons indicate that extrapolation of laboratory test results to field conditions is highly complex. Tests run with organic pollutants can favor certain tolerant species under laboratory conditions; such changes do not, however, alter the application of indexes such as species richness, which is often a robust predictor of field response. Unless the chemical conditions in the laboratory directly simulate the field, the bioassay tests do not adequately predict the field distribution. Resistant populations in the field may also confound the direct application of bioassay results to field conditions, because such resistance may not be developed under the relatively short-term laboratory conditions. Species richness remains an adequate indicator of field conditions when applied to certain well-defined aspects of the field response of indigenous organisms to toxic waste concentrations. Thus, local features of the affected system, the form of the pollutant or pollutants, the duration of the laboratory test, and the index used for prediction are all important variables when one is trying to extrapolate the results of multispecies bioassay tests to field conditions. When carried out in a realistic manner, with adequate field calibration and an understanding of the biological response to the laboratory artifacts, the multispecies leaf-pack bioassay can be used to predict the impact of toxic wastes on aquatic systems. However, such extrapolation is complex and often indirect, so the simplistic use of uncalibrated microcosm bioassays to predict field response to toxic wastes is not feasible.

Acknowledgments

This research was funded by Grant No. CR810554-01-03 from the U.S. Environmental Protection Agency, administered by M. A. Shirazi. Additional administration aid was given by G. C. Woodsum. Statistical and computational assistance was provided by L. E. Wolfe. Species identifications were made by G. L. Ray, W. Karsteter, and J. H. Epler. Field work was carried out by W. H. Clements, M. J. Hollingsworth, and S. B. Holm; C. C. Koenig was in charge of the bioassay program.

References

[1] Cairns, J., Jr., Alexander, M., and Cummings, K. W., *Testing for Effects of Chemicals on Ecosystems*, National Academy Press, Washington, DC, 1981.

[2] Golley, F. B., "What Ecologists Expect from Industry," *Multispecies Toxicity Testing*, J. Cairns, Jr., Ed., Pergamon Press, New York, 1985, pp. 27–35.

[3] Cairns, J., Jr., "Are Single-Species Toxicity Tests Alone Adequate for Estimating Environmental Hazard?" *Hydrobiologica*, Vol. 100, 1983, pp. 47–57.

[4] Hammons, A. S., Giddings, J. M., Suter, G. W., II, and Barnthouse, L. W., *Methods for Ecological Toxicology: A Critical Review of Laboratory Multispecies Tests*, EPA Publication No. 1710, U.S. Environmental Protection Agency, Washington, DC, 1981.

[5] Sloof, W., "The Role of Multispecies Testing in Aquatic Toxicology," *Multispecies Toxicity Testing*, J. Cairns, Jr., Ed., Pergamon Press, New York, 1985, pp. 45–60.

[6] Livingston, R. J., Diaz, R. J., and White, D. C., "Field Validation of Laboratory-Derived Multispecies Aquatic Test Systems," Project Summary, EPA 600/S4-85/039, U.S. Environmental Protection Agency, Washington, DC, 1985, pp. 1–7.

[7] Livingston, R. J. and Meeter, D. A., "Correspondence of Laboratory and Field Results: What are the Criteria for Verification?" *Multispecies Toxicity Testing*, J. Cairns, Jr., Ed., Pergamon Press, New York, 1985, pp. 76–88.

[8] Cairns, J., Jr., and Cherry, D. S., "A Site-Specific Field and Laboratory Evaluation of Fish and Asiatic Clam Population Responses to Coal Fired Power Plant Discharges," *Water Science and Technology*, Vol. 15, 1983, pp. 31–58.

[9] Swartz, R. C., Schults, D. W., Ditsworth, G. R., Deben, W. A., and Cole, F. A., "Sediment Toxicity, Contamination, and Macrobenthic Communities Near a Large Sewage Outfall," *Validation and Predictability of Laboratory Methods for Assessing the Fate and Effects of Contaminants in Aquatic Ecosystems, ASTM STP 865*, T. E. Boyle, Ed., American Society for Testing and Materials, Philadelphia, 1985, pp. 152–175.

[10] Finger, S. E., Little, E. F., Henry, M. G., Fairchild, J. F., and Boyle, T. P., "Comparison of Laboratory and Field Assessment of Fluorine: Part I—Effects of Fluorine on the Survival, Growth, Reproduction, and Behavior of Aquatic Organisms in Laboratory Tests," *Validation and Predictability of Labo-*

ratory Methods for Assessing the Fate and Effects of Contaminants in Aquatic Ecosystems, ASTM STP 865, T. E. Boyle, Ed., American Society for Testing and Materials, Philadelphia, 1985, pp. 120-133.

[*11*] Boyle, T. P., Finger, S. E., Paulson, R. L., and Rabeni, C. F., "Comparison of Laboratory and Field Assessment of Fluorine: Part II—Effects on the Ecological Structure and Function of Experimental Pond Ecosystems," *Validation and Predictability of Laboratory Methods for Assessing the Fate and Effects of Contaminants in Aquatic Ecosystems, ASTM STP 865,* T. E. Boyle, Ed., American Society for Testing and Materials, Philadelphia, 1985, pp. 134-151.

[*12*] Harrass, M. C. and Taub, F. B., "Comparison of Laboratory Microcosms and Field Responses to Copper," *Validation and Predictability of Laboratory Methods for Assessing the Fate and Effects of Contaminants in Aquatic Ecosystems, ASTM STP 865,* T. E. Boyle, Ed., American Society for Testing and Materials, Philadelphia, 1985, pp. 57-74.

[*13*] Livingston, R. J., "Field Verification of Bioassay Results at Toxic Waste Sites in Three Southeastern Drainage Systems," Project Summary, U.S. Environmental Protection Agency, Washington, DC, in press.

[*14*] MacArthur, R. H. and Wilson, E. O., "An Equilibrium Theory of Insular Zoogeography," *Evolution,* Vol. 17, 1963, pp. 373-387.

Hans Blanck,[1] *Sten-Åke Wängberg,*[1] *and Sverker Molander*[1]

Pollution-Induced Community Tolerance—A New Ecotoxicological Tool

REFERENCE: Blanck, H., Wängberg, S.-Å., and Molander, S., **"Pollution-Induced Community Tolerance—A New Ecotoxicological Tool,"** *Functional Testing of Aquatic Biota for Estimating Hazards of Chemicals, ASTM STP 988,* J. Cairns, Jr., and J. R. Pratt, Eds., American Society for Testing and Materials, Philadelphia, 1988, pp. 219-230.

ABSTRACT: The authors hypothesize that pollution-induced community tolerance (PICT) is direct evidence that a community is disturbed by a pollutant and, furthermore, that the agent or agents causing the effects can be identified, because induced tolerance will be observed only for those compounds that have exerted selection pressure on the community. A similar concept has been formulated for population tolerance, but we suggest that the concept is more useful when applied at the community level. Our examination of some crucial points behind PICT, using arsenate and periphyton communities, demonstrates that (1) the tolerance increase can be determined in short-term photosynthesis experiments, (2) PICT correctly indicates changes in species composition and net production, and (3) the selection pressure of arsenate is specific—that is, a tolerance increase for one compound is not followed by co-tolerance to other compounds unless they are closely related chemically or in their mode of action. These results support the idea of using PICT as an ecotoxicological tool. If further corroborated, PICT can be applied in both laboratory and field studies. PICT might be used retrospectively in the field to detect minor disturbances and to identify the causing agent. In a laboratory test system, PICT can be used to estimate the no-effect concentration for the community under study. The authors argue that test systems based on PICT will be sensitive and yield ecologically relevant information.

KEY WORDS: hazard evaluation, arsenate, aquatic toxicology, co-tolerance, ecotoxicology, multispecies tests, periphyton, photosynthesis, phytoplankton, toxicant tolerance

Ideally, ecotoxicological tools should be simple, fast, and reproducible; should generate ecologically relevant information; and should be able to establish causal relationships firmly. Laboratory and field investigations are complementary in meeting these criteria, and no single technique or methodology has so far combined the merits of both. Not only is it difficult to extrapolate the response of a short-term test system at a low level of biological organization to longer-term impact on ecosystems [1], but it is also difficult at present to distinguish pollution-induced changes in ecosystems from those due to other causes [2,3]. Ecotoxicology would benefit if ecologically realistic experimental systems with high test potential were available, and if, in field studies, one could also detect minor perturbations with certainty and identify their causes.

Pollution-induced community tolerance (PICT) is proposed as a new ecotoxicological tool that combines some of the advantages of both approaches. PICT is assessed by comparing the tolerance of, for example, photosynthesis in algal community samples taken from sites representing a gradient in pollutant stress. We hypothesize that PICT is direct evidence that a community is affected by toxicants present in an ecosystem, and that the agents actually affecting the biota can be identified because PICT will be observed only for those compounds that exert

---

[1]Research associate and Ph.D. students, respectively, Department of Plant Physiology, University of Göteborg, S-413 19 Göteborg, Sweden.

selection pressure in the community. PICT is potentially applicable in both laboratory and field studies.

A similar concept focusing on the tolerance increase in separate populations was first formulated by Luoma [4]. We suggest that some advantages are gained when the concept is applied at the community level.

In this paper we present the rationale behind the PICT approach, discuss its potential applications, and examine some prerequisites and evidence with bearing on the use of PICT as an ecotoxicological tool.

Rationale

A toxicant in a polluted environment will affect biota by stressing or eliminating sensitive individuals or populations of organisms. If the stress is strong enough, the exposed organisms will be unable to survive or reproduce. The chemical will exert a selection pressure on the organisms, and their fitness will be affected. In such a case the organisms will respond to the stress by either avoiding it, adapting to it, or dying. Each of these responses will lead to the replacement of sensitive individuals by tolerant ones and the overall tolerance of the surviving populations will increase in that location. Furthermore, the extinction of any population unable to adapt will add to the community tolerance. Thus, community tolerance can increase by three different mechanisms: (1) the physiological adaptation of individuals, (2) selection of tolerant genotypes within a population, and (3) replacement of sensitive species by tolerant ones in a community. Measurements of PICT will usually not discriminate between these alternatives.

We postulate that chemicals that do not exert a selection pressure cannot cause any significant biological disturbance in an ecosystem, since they are unable to restructure the communities or to change the genotype distribution in the different populations. Thus, only those effects recognized by the biota as significant for their survival in the environment will be expressed in PICT.

The induction of tolerance in a community is not a remedy for toxic effects but should be considered evidence that a toxicant is a structuring factor in the community. PICT thus indicates a disturbed community.

Advantages of PICT in Comparison with the Population Tolerance Approach

In the population tolerance approach [4], it was proposed that induced tolerance to a toxicant in a population is direct evidence that the concentration of the toxicant in the environment of the tolerant population is sufficient to elicit biological effects. The PICT approach is very similar to this but is not restricted to tolerance of single populations. With the population tolerance approach it has to be *assumed* that the presence of a tolerant population indicates that other species may have been affected. This assumption is not needed when using PICT, because any change in the community (such as the exclusion of sensitive species) due to the selection pressure will contribute to the observed tolerance increase. Every effect is therefore accounted for directly.

PICT operates at the community level and employs naturally derived community samples and can therefore, with only minor modification, be used in a variety of ecosystems. It is more generally applicable, as it does not rely on any particular test species. When the population tolerance approach is used, the choice of test species is critical. This restriction to separate species has the implication that one must by chance find those that are tolerant enough to survive the stress but sensitive enough to recognize and adapt to it. This is difficult and very similar to the problem of finding the representative or the most sensitive test species when designing conventional one-species bioassays [5–7]. The ability of the population tolerance approach to detect an impact in a receiving body of water is therefore occasional and will depend on whether the appropriate test species has been chosen. In contrast, PICT will respond as soon as any part of the community is excluded by the selection pressure, a property which also contributes to the sensitivity of PICT.

The tolerance increase in a population could be due to phenotypic adaptation or to selection for tolerant genotypes, and these two mechanisms will determine the possible extent of the tolerance increase. The principal difference is that, in addition to these, PICT also makes use of the interspecific variation in tolerance. The potential tolerance increase is therefore likely to be greater in the community and more easily detected.

The advantages of PICT thus include its more direct way of indicating damage to the ecosystem, its more general applicability to pollution situations, and its higher sensitivity.

Prerequisites, Evidence, and Problems

The successful application of PICT as an ecotoxicological tool along the lines just proposed has the following requirements:

1. Tolerance can be measured in a meaningful way in short-term test systems.
2. PICT can be detected and distinguished from other causes of variation in community tolerance.
3. PICT is associated with significant effects on the community as measured by conventional methodologies.
4. The specificity of the selection pressure is high, or understandable patterns in co-tolerance can be found.

The Occurrence and Measurement of Tolerance in Populations and Communities

It is well established that populations of organisms do respond to chemical stress as a selection pressure [8]. Although phenotypic changes in individuals are possible [9,10], generally the increased tolerance in a population seems to be due to a selection of tolerant genotypes [4,11]. A genetic variability is a prerequisite for metal tolerance in angiosperms in the terrestrial environment [12].

The most ecologically relevant measure of tolerance is the ability of a population to survive and persist in a polluted environment. With that definition of tolerance, however, it would not be possible to quantify any difference in tolerance between two locations—at least not in a way that could be used to relate the presence of a toxicant to an impact on the system. To be able to do that, we must define tolerance operationally as that tolerance which can be measured in some kind of ecotoxicological test system. Only in this manner can cause and effect be firmly connected. The approach is, however, somewhat indirect in that it will only detect tolerance due to mechanisms on the physiological level within the organism. Gray [2], following Grime [13], pointed out the significance of various life-cycle strategies in conveying tolerance to marine benthic invertebrates. Tolerance due to this type of strategy will not be detected in a PICT analysis using the operational definition just given.

In aquatic environments subject to pollution stress, increased tolerance has been found in populations of bacteria [14], algae [15–22], and aquatic animals such as isopods [23,24], copepods [25], chironomids [26], polychaetes [27–30], gastropods [25], and fish [31–33].

While it is well established that aquatic species can increase their tolerance to toxicants, the evidence for increased community tolerance as a result of toxicant stress is scarce and mostly indirect. In the controlled ecosystem pollution experiment (CEPEX) program, copper added to enclosed pelagic communities led to the replacement of centric diatoms by microflagellates and pennate diatoms [34]. Photosynthesis of the new plankton communities had a higher copper tolerance [35], which was related to the preexposure concentration. Sanders et al. [36] in a similar experiment added high concentrations of copper (25 to 40 μg/L) to a phytoplankton community. A dramatic change in species composition led to an extensive bloom of an otherwise uncommon diatom species, *Amphiprora paludosa* var. *hyalina*. Copper obviously affected the competitive ability of the phytoplankton, but it was not possible to attribute that to a physiological tolerance of the *Amphiprora* species since it did not grow in pure cultures. Its copper tolerance was, however, clearly indicated by its persistence in the copper-loaded system.

Marine periphyton communities established under arsenate stress increased their community tolerance (Fig. 1).[2] The tolerance increase was related to the severity of the stress under which the communities were established, in a dose/response-like manner. Similarly, freshwater periphyton and phytoplankton in limnocorrals increased their community tolerance after exposure to arsenate.[3]

In these cases it was possible to record changes in the algal communities by measuring the arsenate tolerance of photosynthesis. It is, however, still an open question whether photosynthesis generally is a suitable test parameter to indicate PICT in algal communities. Therefore, the test parameters used to assess PICT must be chosen with care when studying other toxicants.

When the population tolerance approach was first formulated [4], no means of quantifying the tolerance increase was suggested, but in a later study by Luoma and co-workers [25] of the bivalve *Macoma baltica* and the copepod *Acartia clausi*, lethality bioassays were employed. Various workers applying the population tolerance concept have generally used common short-term bioassays to assess the tolerance differences. Lethality tests with marine animals have also been used in combination with the analysis of allelic isozyme patterns [37].

Detection of increased community tolerance will depend on the variability in tolerance in communities not exposed to the pollutants. This has not been systematically investigated. Gächter [38] reported fairly large variation in phytoplankton sensitivity to heavy metals over the year [the ratio of maximum effective concentration (EC_{max}) to minimum effective concentration (EC_{min}) varied from 9 to 50], but samples taken during the same season were much less variable—the mean EC_{max}/EC_{min} value was 3.6 for 55 observations. Periphyton communities from various streams showed a relatively small variation in sensitivity to aliphatic amines and a textile industry effluent [39]; with very few exceptions, EC_{max}/EC_{min} was less than 3.5. Seasonal variation at the same location was even less. A marine periphyton community sampled in different years had a fairly constant arsenate tolerance (the EC_{max}/EC_{min} was less than 6)[4] unless exposed to arsenate stress or periods with high phosphate concentrations. Although there is some uncertainty as to the generality of these findings, different periphyton communities appear to be similar in their tolerance even when they differ substantially in biomass and species composition.

Validation of PICT

Before the population tolerance and PICT concepts can be used as valid evidence of toxicant impact on ecosystems, it must be shown that there is a relationship between the induced tolerance and gross changes in the communities as detected by conventional means.

The mere presence of induced tolerance in polluted environments is not sufficient to validate the two concepts. Instead, it must be demonstrated that the same toxicant caused both the tolerance increase and other effects. This can be done conclusively only in experimental systems or in very simple and well-known pollution situations with only one stress factor. It is of course pointless to make a validation effort in an environment where one runs the risk of relating the increase in tolerance to one compound to the gross effects of another. If similar dose-response relationships are not obtained for the induced tolerance and other ecologically relevant changes

[2]Blanck and Wängberg, "Induced Community Tolerance in Marine Periphyton Established Under Arsenate Stress," *Canadian Journal of Fisheries and Aquatic Sciences*, in press.

[3]Wängberg et al., "Arsenate Tolerance in Freshwater Periphyton and Phytoplankton Established Under Arsenate Stress in Limnocorrals," in preparation.

[4]Wängberg and Blanck, "Arsenate Tolerance in Marine Periphyton Communities Under Different Nutrient Regimes," in preparation.

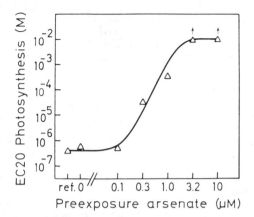

FIG. 1—*Community tolerance to arsenate in marine periphyton communities established under arsenate stress. Tolerance was measured in short-term photosynthesis tests using three-week-old communities. The reference value is for a reference community established in the Gullmar fjord, Sweden, while all the other communities were established in fjord water indoors. EC_{20} is the effective concentration that reduces the photosynthetic activity by 20%. This illustration is used with permission of the* Canadian Journal of Fisheries and Aquatic Sciences.

in the community, PICT might generate false positive or false negative indications of pollutant stress.

The population tolerance concept has been examined in the sense that tolerance is induced in populations exposed to toxicants both in the laboratory and in the field [25]. It has also been applied as additional evidence of toxicant impact on a population [33,40]. The suggested extrapolation from induced tolerance in a population to impact on a community has, however, not been challenged.

Marine periphyton communities experimentally stressed by arsenate were used to examine whether PICT shows the same concentration dependence as do other long-term changes of the algal communities.[5] Effects on biomass production and species composition occurred at arsenate concentrations between 0.2 and 0.8 μM (Table 1). Comparison of these concentrations with the arsenate level giving rise to PICT (Fig. 1) demonstrates that there is a close relationship between the tolerance increase which occurs at concentrations between 0.1 and 0.3 μM and gross effects on community properties. To our knowledge this is the first evidence that induced tolerance can be used as an indicator of toxicant impact at the community level.

This effect, however, was shown under constant arsenate stress only, and the usefulness of PICT in environments with variable loading of pollutants remains to be investigated.

Co-Tolerance in Populations and Communities

The anticipated specificity of the selection pressure led Luoma [4] to suggest that induced tolerance would be observed only for those compounds eliciting biological effects in a population. The stress from one toxicant must not result in an increase in tolerance to other compounds to which the organisms had not been exposed at levels high enough to be affected. In other words, for this concept to hold strictly, co-tolerance must not be a common phenomenon.

Whether co-tolerance should be expected to occur or not will depend on the means of conferring tolerance and on the tolerance mechanisms. There is considerable evidence that bacteria

---

[5]Blanck and Wängberg, "The Validity of an Ecotoxicological Test System: Short-Term and Long-Term Effects of Arsenate on Marine Periphyton Communities in Laboratory Systems," *Canadian Journal of Fisheries and Aquatic Sciences,* in press.

TABLE 1—*The tolerance of marine periphyton communities to arsenate, recorded with various short- and long-term test parameters.*[a,b]

Test Parameter	EC_{20}, μM^b
Short-term effects	
Reference community	0.4
Control tank community	0.6
Long-term effects	
Species composition	0.2
Chlorophyll *a* accumulation	0.3
Total carbon accumulation	0.5
Total nitrogen accumulation	0.8
Photosynthesis per unit area	0.3

[a]All measurements were made on three-week-old communities.
[b]EC_{20} is the effective concentration that reduces an activity or a factor by 20%.

often carry multiple resistance to various metals and antibiotics on the same plasmid [41,14]. Any transfer of such a plasmid would lead to co-tolerance also when the tolerance mechanisms are entirely different.

The presence of detoxifying mechanisms where a broad range of compounds can be eliminated by degradation, such as by the mixed-function oxygenase system [42], or by binding to certain molecules, such as metallothioneins [43,44] in fish, will also confer co-tolerance. A tolerance strategy involving the modification of the target site would, however, be more likely to be specific. Luoma [4] argued that such a strategy has been a dominant one in the metal tolerance of grasses and in the pesticide tolerance of insects. Although the mechanisms leading to tolerance in aquatic organisms are not well known, it appears that the appropriate question is not whether co-tolerance occurs or not, but whether it is a common phenomenon and understandable patterns of co-tolerance can be identified.

Metal tolerance of algae present in metal-contaminated areas is well documented, and multiple tolerance, that is, elevated tolerance to several metals, is often found [15,18]. However, in field studies most claims of co-tolerance are inconclusive, since one cannot with certainty exclude the possibility that the algae were simultaneously exposed to several toxicants in the environment. In field studies it is therefore very difficult to discriminate between multiple tolerance and co-tolerance. Only exposure of organisms to one selection pressure at a time, under controlled conditions can help to determine whether co-tolerance occurs or not. Laboratory investigations of metal co-tolerance, have, however, yielded conflicting results. Stokes [45] selected algae for copper tolerance and found co-tolerance for nickel but not for cobalt or cadmium. Whitton and Shehata [46] found little evidence of co-tolerance in *Anacystis nidulans* using the same metals; instead, a decreased tolerance was often observed. Also in angiosperms the co-tolerance patterns are variable [47] and it appears that plants have the ability to use several different tolerance mechanisms, of which some will lead to co-tolerance while others will not.

A similar situation is found in tolerance to the herbicide atrazine. The target site of atrazine (and many other herbicides) is on a membrane-bound protein in the chloroplast, where these herbicides affect photosynthetic electron transport [48]. Tolerance is acquired by modification of the target site. Co-tolerance occurs for herbicides that affect the same site of this protein. A complicated co-tolerance pattern has been observed, however, and the investigators presented a model suggesting that co-tolerance will occur for two herbicides if they share a common domain of the herbicide binding site. A similar model was proposed for herbicide-tolerant algae [49]. Exchange of single amino acids of the 32-kilodalton herbicide-binding protein is sufficient to produce atrazine tolerance in algae and cyanobacteria [50,51]. Whether or not co-tolerance with diuron occurs depends on which amino acid is exchanged [50]. This is a case in which co-tolerance patterns have provided very precise information on the mode of action of the toxicants.

Another example of co-tolerance is found for the herbicide paraquat and the sulfite ion [52]. Preexposure of *Chlorella sorokiniana* to sulfite, phenotypically induced, increased activity of the superoxide dismutase, which rendered the alga more tolerant to paraquat. The results suggest that there is a similar mode of action for both toxicants, presumably involving production of the superoxide radical.

These examples of documented co-tolerance suggest that it should be looked for among groups of chemicals that either affect the same target in the same manner, produce a common intermediate, or are chemically very similar.

In our examination of arsenate-tolerant algal communities we looked for co-tolerance among toxicants chemically or mechanistically related to arsenate.[6] Arsenate inhibits photosynthesis by uncoupling of the photophosphorylation [53], and therefore several uncouplers [carbonyl cyanide *m*-chlorophenylhydrazone (CCCP), a mixture of highly branched primary aliphatic amines (Primene JMT), and sodium thiophosphate] were tested for co-tolerance. Of these, only thiophosphate has the same mode of action as arsenate; they both are phosphate analogs and are supposed to uncouple photophosphorylation by forming unstable analogs of adenosine triphosphate (ATP), which immediately reform the adenosine diphosphate (ADP) [53,54]. Arsenite, which is chemically related to arsenate and is proposed as an intermediate in the algal metabolism of arsenate [55], was also tested. The inhibitor of photosynthetic electron transport diuron [3-(3′,4′-dichlorphenyl)-1,1-dimethylurea (DCMU)] was included as a control with no known similarity to arsenate. Although the arsenate tolerance was increased 20 000-fold, no significant co-tolerance occurred except for the thiophosphate ion (Fig. 2).

Thus, it appears that a specificity similar to that observed in the biochemical studies just described is maintained also when the selection pressure is applied at the community level. Co-tolerance occurs only for compounds closely related, either chemically or in their mode of action.

Potential Applications of PICT

PICT may be used both predictively in laboratory studies and retrospectively in field studies. PICT is assessed by comparing tolerance levels in community samples from sites representing a gradient in pollutant stress (Fig. 3). These sites can be locations along a transect in a polluted environment or a set of simple aquaria or mesocosms in an experimental investigation of toxicants. Community tolerance is operationally defined as the tolerance of some test parameter in relation to functions or processes in the community, for example, photosynthesis or respiration. Toxicants that are suspected of having effects in the system are assessed for PICT. An observed tolerance gradient along a pollution gradient or a toxicant concentration series is evidence that the toxicant has exerted a selection pressure on the community and thus produced biological effects.

PICT in Test Systems

In a test system, PICT may be used to estimate the no-effect concentration (NEC) for the community under study. Communities are established in a set of simple aquaria or mesocosms under various degrees of toxicant stress. When the toxicant concentration is high enough to exert a selection pressure, sensitive components of the community are unable to compete, and are replaced by more tolerant components. These components may be species, genotypes, or

[6]Blanck and Wängberg, "Pattern of Co-Tolerance in Marine Periphyton Communities Established Under Arsenate Stress," in preparation.

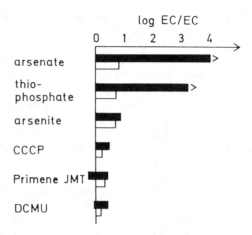

FIG. 2—*Co-tolerance pattern in arsenate-tolerant marine periphyton communities. The tolerance increase for various toxicants is given as the effective concentration ratio (EC/EC) between arsenate-tolerant and arsenate-sensitive communities* (black bars). *The observed variation in tolerance among the arsenate-sensitive communities is given for comparison* (white bars). *The photosynthesis tolerance was determined in short-term experiments.*

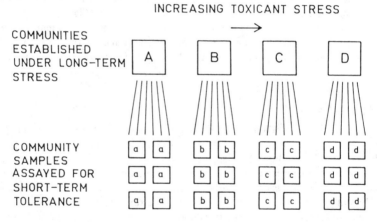

FIG. 3—*The principle of PICT assessment. Communities are established (preexposed) in one concentration gradient of the toxicant. Subsamples are taken and exposed to a new concentration series, and their tolerance is assayed in a short-term toxicity test. The tolerance levels thus determined for Communities A through D are then compared and related to the exposure level, as illustrated in Figs. 4 and 5.*

phenotypes. The overall tolerance of such communities will have increased in comparison with that of communities which do not recognize a selection pressure from the toxicant. The highest level of toxicant stress that does not induce a tolerance increase is the NEC. The principle is illustrated in Fig. 4.

Any community-level test system that employs naturally derived community samples, that allows sufficient replication, and that uses metabolic activity (or another functional property) as a test parameter can probably be used in PICT studies. However, it has only been applied so far

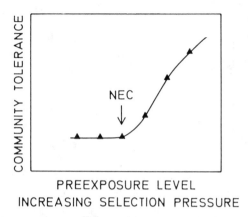

PREEXPOSURE LEVEL
INCREASING SELECTION PRESSURE

FIG. 4—*The use of PICT in experimental studies for determination of the no-effect concentration (NEC) in communities under toxicant stress. NEC is the highest concentration that does not induce community tolerance. At higher concentrations the toxicants exert a selection pressure on the community, and sensitive components are excluded, which leads to a tolerance increase.*

to periphyton and phytoplankton communities, with the photosynthesis tolerance being estimated in short-term experiments.[7]

We believe that test systems based on PICT are sensitive since the tolerance increase starts with the exclusion of the most sensitive components of the community. Thus, it is an inherent property of PICT that it scans the community for its most sensitive components. PICT should also yield ecologically relevant information, because only those effects of vital importance to the organism in its association with other members of the community will be expressed in PICT.

PICT in the Field

By measuring tolerance gradients among community samples in a receiving body of water, one might establish whether a suspected pollutant has affected the community or not. Also, the influenced area—that is, how far from the emission point a certain toxicant has affected the community—can be determined. When no further decrease in tolerance is observed with distance from the outlet, the outer limit of the area of influence has been reached for the community and toxicant under study. Under complex emission and mixing situations, distance cannot be directly related to the effluent concentration, and the actual exposure levels will have to be determined.

Ideally, a gradient in tolerance will be observed only for those toxicants that have affected the community, while the tolerance to trace contaminants that have not affected the community will be similar at all stations along the pollution gradient (Fig. 5). However, this is strictly true only when co-tolerance does not occur and when tolerance is not increased as a result of causes other than the toxicant stress.

The approach further requires that communities not exposed to the selection pressure be similar in their tolerance so that a tolerance baseline can be determined. Unpolluted reference locations outside the investigated area are needed.

[7]Blanck and Wängberg, "Induced Community Tolerance in Marine Periphyton Established Under Arsenate Stress," *Canadian Journal of Fisheries and Aquatic Sciences,* in press; and Wängberg et al., "Arsenate Tolerance in Freshwater Periphyton and Phytoplankton Established Under Arsenate Stress in Limnocorrals," in preparation.

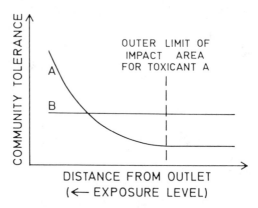

FIG. 5—*The use of PICT in field studies to determine the impact area of pollutants and to identify the toxicant causing the impact. Tolerance gradients will occur only for toxicants that have exerted selection pressure on the community. Toxicant A has affected the community, while toxicant B has not.*

Special Applications of PICT

Distinguishing Between Primary and Secondary Effects—A toxicant that does not have its primary effect on algae may still have the potential to cause secondary effects in the algal community because of, for example, altered grazing pressure. With conventional methods it is usually very difficult to discriminate between primary and secondary effects. The PICT method may have this capacity. If the selection pressures of toxicants are specific, then PICT would respond only to primary effects, that is, when the toxicant is exerting selection pressure directly on the algal members of the community. Secondary effects, for example altered grazing pressure or competition with bacteria for nutrients, would not affect PICT in algae.

Estimation of the Toxicant Mode of Action at the Community Level—If co-tolerance is frequent in polluted environments a proper identification of the selection pressure cannot be done; however, if further patterns of co-tolerance emerge, it might be possible to delimit the suspected pollutants to a handful of chemicals that are closely related. If the toxicant for which PICT was recorded is not causing the impact, it is most probably a compound with similar characteristics.

It might even be possible to estimate the mode of action of an unidentified compound present in an industrial effluent if it exerts a selection pressure. To do this, community samples should be assayed for co-tolerance to a set of toxicants with well-defined targets and modes of action. Various herbicides and metabolic inhibitors could be suitable tools to study algal communities. If co-tolerance patterns are identifiable, induced tolerance to one of the standard inhibitors would imply a similar mode of action of the compound causing the effect in the community. Some evidence for this type of co-tolerance pattern in periphyton communities was presented in the section earlier in this paper on co-tolerance in populations and communities.

Concluding Remarks

We have presented evidence that a toxicant that exerts selection pressure on a periphyton community will restructure it and give rise to pollution-induced community tolerance (PICT). The tolerance increase was assessed using a short-term test system measuring the inhibition of periphyton photosynthesis. PICT correctly indicated disturbance of the community. Furthermore, the tolerance increase was fairly specific, with insignificant co-tolerance except for toxicants that are closely related. These lines of evidence concerning the effects of arsenate are the first empirical data in support of PICT as an ecotoxicological tool. Although the PICT concept is supported by our results on arsenate and algal communities, its general applicability to other toxicants and communities remains to be investigated. Further work is also needed on other aspects of PICT, the most important being its response to variable stress and mixtures of toxicants and the identification of co-tolerance patterns. If further corroborated the concept will find important applications in ecotoxicology.

Gray [2] argued that conventional means of detecting effects on community structure are both insensitive and not capable of discriminating clearly between affected and unaffected communities. Looking at the log-normal distribution of individuals among species was a more sensitive method but that too was unable to relate cause and effect specifically in field studies. Proper assessment of PICT in the field may have that capacity and the ability to detect minor disturbances.

PICT differs from most other test parameters by its direct connection to central ecological concepts such as fitness and natural selection. It is also coupled to the realized and not the fundamental niche of an organism, since it measures the sensitivity of organisms in their association with other community members. It is a short-term, functional test but indicative of structural changes in the community. In this manner PICT combines ecological relevance with a high test potential. If the assumption that only toxicants that exert selection pressure can restructure and damage a community is correct, then a direct measurement of the ecologically relevant no-effect concentration for a given community can be implemented. Such measurements might form an operational basis for a new definition of environmental hazards as those contaminants that constitute anthropogenic selection pressures and thereby act as structuring factors in the ecosystem.

References

[1] National Research Council, *Testing for Effects of Chemicals on Ecosystems*, National Academy Press, Washington, DC, 1981, p. 103.

[2] Gray, J. S., *Philosophical Transactions of the Royal Society of London, Series B: Biological Sciences*, Vol. 286, 1979, pp. 545-561.

[3] Livingston, R. J., *Ecological Stress and the New York Bight: Science and Management*, Estuarine Research Federation, New York, 1982, pp. 605-620.

[4] Luoma, S. N., *Journal of the Fisheries Research Board of Canada*, Vol. 34, 1977, pp. 436-439.

[5] Slooff, W., Canton, J. H., and Hermens, J. L. M., *Aquatic Toxicology*, Vol. 4, 1983, pp. 113-128.

[6] Blanck, H., *Ecological Bulletins*, Vol. 36, 1984, pp. 107-119.

[7] Blanck, H., Wallin, G., and Wängberg, S.-A., *Ecotoxicology and Environmental Safety*, Vol. 8, 1984, pp. 339-351.

[8] Evered, D. and Collins, G. M., Eds., *Origins and Development of Adaptation*, Ciba Foundation Symposium 102, Pitman Press, London, 1983.

[9] LeBlanc, G. A., *Environmental Pollution, Series A*, Vol. 27, 1982, pp. 309-322.

[10] Duncan, D. A. and Klaverkamp, K., *Canadian Journal of Fisheries and Aquatic Sciences*, Vol. 40, 1983, pp. 128-138.

[11] Beardmore, J. A., Barker, C. J., Battaglia, B., Berry, R. J., Longwell, A. C., Payne, J. F., and Rosenfield, A., *Rapports et Proces-Verbaux des Reunions Conseil International pour l'Exploration de la Mer*, Vol. 179, 1980, pp. 299-305.

[12] Bradshaw, A. D. in *Symposium on Origins and Development of Adaptation*, Ciba Foundation Symposium 102, Pitman Press London, 1983, pp. 4-19.

[13] Grime, J. P., *The American Naturalist*, Vol. 111, 1977, pp. 1169-1195.

[14] Trevors, J. T., Oddie, K. M., and Belliveau, B. H., *FEMS Microbiology Reviews*, Vol. 32, 1985, pp. 39-54.

[15] Stokes, P. M., Hutchinson, T. C., and Krauter, K., *Canadian Journal of Botany*, Vol. 51, 1973, pp. 2155-2168.

[16] Jensen, A. and Rystad, B., *Journal of Experimental Marine Biology and Ecology*, Vol. 15, 1974, pp. 145-157.

[17] Harding, J. P. C. and Whitton, B. A., *British Phycological Journal*, Vol. 11, December 1976, pp. 417-426.

[18] Foster, P. L., *Freshwater Biology*, Vol. 12, 1982, pp. 41-61.

[19] Murphy, L. S., Guillard, R. R. L., and Gavis, J., *Ecological Stress and the New York Bight: Science and Management*, Estuarine Research Federation, New York, 1982, pp. 401-413.

[20] Reed, R. H. and Moffat, L., *Journal of Experimental Marine Biology and Ecology,* Vol. 69, 1983, pp. 85-103.

[21] Stokes, P. M. in *Progress in Phycological Research,* F. E. Round and D. J. Chapman, Eds., Elsevier, New York, 1983, Chapter 3, pp. 87-112.

[22] Cosper, E. M., Wurster, C. F., and Rowland, R. G., *Marine Environmental Research,* Vol. 12, 1984, pp. 209-223.

[23] Brown, B. E., *Water Research,* Vol. 10, 1976, pp. 555-559.

[24] Fraser, J., Parkin, D. T., and Verspoor, E., *Water Research,* Vol. 12, 1978, pp. 637-641.

[25] Luoma, S. N., Cain, D. J., Ho, K., and Hutchinson, A., *Marine Environmental Research,* Vol. 10, 1983, pp. 209-222.

[26] Wentsel, R., McIntosh, A., and Atchison, G. in *Bulletin of Environmental Contamination and Toxicology,* Vol. 20, 1978, pp. 451-455.

[27] Bryan, G. W. in *Pollution and Physiology of Marine Organisms,* F. J. Vernberg and W. B. Vernberg, Eds., Academic Press, New York, 1974, pp. 123-135.

[28] Bryan, G. W. and Hummerstone, L. G., *Journal of the Marine Biological Association of the United Kingdom,* Vol. 51, 1971, pp. 845-863.

[29] Bryan, G. W. and Hummerstone, L. J., *Journal of the Marine Biological Association of the United Kingdom,* Vol. 53, 1973, pp. 839-857.

[30] Bryan, G. W. and Hummerstone, L. J., *Journal of the Marine Biological Association of the United Kingdom,* Vol. 53, 1973, pp. 859-872.

[31] Ferguson, D. E., Culley, D. D., Cotton, W. D., and Dodds, R. P., *Bioscience,* Vol. 14, No. 11, 1964, pp. 43-44.

[32] Grant, B. F., *Bulletin of Environmental Contamination and Toxicology,* Vol. 15, 1976, pp. 283-290.

[33] Weis, J. S. and Weis, P., *Marine Environmental Research,* Vol. 13, 1984, pp. 231-245.

[34] Thomas, W. H. and Seibert, D. L. R., *Bulletin of Marine Science,* Vol. 27, No. 1, 1977, pp. 23-33.

[35] Harrison, W. G., Eppley, R. W., and Renger, E. H., *Bulletin of Marine Science,* Vol. 27, 1977, pp. 44-57.

[36] Sanders, J. G., Batchelder, J. H., and Ryther, J. H., *Botanica Marina,* Vol. 24, 1981, pp. 39-41.

[37] Nevo, E., Lavie, B., and Ben-Shlomo, R. in *Isozymes: Current Topics in Biological and Medical Research: Genetics and Evolution,* Vol. 10, 1983, pp. 69-92.

[38] Gächter, R., *Schweizerische Zeitschrift fuer Hydrologie,* Vol. 38, 1976, pp. 97-119.

[39] Blanck, H., *Hydrobiologia,* Vol. 124, 1985, pp. 251-261.

[40] Fisher, N. S. and Frood, D., *Marine Biology,* Vol. 59, 1980, pp. 85-93.

[41] Foster, T. J., *Microbiological Reviews,* Vol. 47, No. 3, September 1983, pp. 361-409.

[42] Payne, J. F., *Ecotoxicological Testing for The Marine Environment,* Vol. 1, G. Persoone, E. Jaspers, and C. Claus, Eds., State University of Ghent, Ghent, Belgium, 1984, pp. 625-655.

[43] Klaverkamp, J. F., McDonald, W. A., Duncan, D. A., and Wagemann, R. in *Contaminant Effects on Fisheries,* V. W. Cairns, P. V. Hodson, and J. O. Nriagu, Eds., Wiley, New York, 1984, pp. 99-113.

[44] Kito, H., Tazawa, T., Ose, Y., Sato, T., and Ishikawa, T., *Comparative Biochemistry and Physiology,* Vol. 73C, No. 1, 1982, pp. 135-139.

[45] Stokes, P. M., *Journal of Plant Nutrition,* Vol. 3, Nos. 1-4, 1981, pp. 667-678.

[46] Whitton, B. A. and Shehata, F. H. A., *Environmental Pollution, Series A: Ecological and Biological,* Vol. 27, 1982, pp. 275-281.

[47] Woolhouse, H. W. in *Physiological Plant Ecology,* Vol. III, *Encyclopedia of Plant Physiology, New Series,* O. L. Lange, P. S. Nobel, C. B. Osmond, and H. Ziegler, Eds., Springer-Verlag, Berlin, 1983, Chapter 7, pp. 245-300.

[48] Arntzen, C. J., Pfister, K., and Steinback, K. E. in *Herbicide Resistance in Plants,* H. M. Le Baron, Ed., Wiley, New York, 1982, Chapter 10, pp. 185-214.

[49] Böger, P., Sandmann, G., and Miller, R., *Photosynthesis Research,* Vol. 2, 1981, pp. 61-74.

[50] Erickson, J. M., Rahire, M., and Rochaix, J.-D., *Science,* Vol. 228, April 1985, pp. 204-207.

[51] Golden, S. S. and Haselkorn, R., *Science,* Vol. 229, September 1985, pp. 1104-1107.

[52] Rabinowitch, H. D. and Fridovich, I., *Planta,* Vol. 164, 1985, pp. 524-528.

[53] Avron, M. and Jagendorf, A. T., *Journal of Biological Chemistry,* Vol. 234, No. 4, 1959, pp. 967-972.

[54] Avron, M. and Shavit, N., *Biochimica et Biophysica Acta,* Vol. 109, 1965, pp. 317-331.

[55] Andreae, M. O. in *Arsenic: Industrial, Biomedical and Environmental Perspectives,* W. H. Lederer and R. J. Fensterheim, Eds., Van Nostrand Reinhold, 1983, pp. 378-392.

Summary

At the outset of the development of a major new branch of environmental assessment, only a few things can be demonstrated: (1) that a methodology producing a totally different kind of important information can aid in management decisions, in this case, for hazard assessment; (2) that this methodology is not only an idea but a working hypothesis with techniques already in place; (3) that it does not have a high site specificity—it will work in more than one part of the world; (4) that it is quantitative and, therefore, as easily utilized as the more conventional information, although it is qualitatively different; and (5) that there are some case history illustrations of how the method has been used.

After demonstrating that a new field has emerged which might be of value to readers of ASTM publications, it then remains for the following steps to occur: (1) collection of evidence on other methods not included in this volume; (2) wider use of the methods that seem the most suitable for management and regulatory decisions; (3) identification of the strengths, defects, and operating conditions for the methods; (4) organization of a task force or task forces to prepare one or more of these methods as candidate standard methods; and (5) processing of methods in the usual fashion. Undoubtedly, additional summary volumes will be necessary at various stages of this process and even after some of the functional tests have become well-established standard methods. The main purpose of this specific special technical publication is to raise the level of awareness about the potential future utilization of such methods and to act as a focal point for further efforts. It is by no means a definitive volume but rather a catalyst for further action.

John Cairns, Jr.

University Center for Environmental and Hazardous Materials Studies, Virginia Polytechnic Institute and State University, Blacksburg, VA 24061; symposium chairman and editor.

James R. Pratt

School of Forest Resources, Pennsylvania State University, University Park, PA 16802; editor.

Author Index

Subject Index